American Forage & Grassland Council Annual Conference 2016

Baton Rouge, Louisiana, USA
10 - 12 January 2016

ISBN: 978-1-5108-3752-2

Printed from e-media with permission by:

Curran Associates, Inc.
57 Morehouse Lane
Red Hook, NY 12571

Some format issues inherent in the e-media version may also appear in this print version.

Printed by Curran Associates, Inc. (2017)

American Forage and Grassland Council (AFGC)
PO Box 867
Berea, KY 40403
USA

Phone: 800.944.2342

www.afgc.org

Additional copies of this publication are available from:

Curran Associates, Inc.
57 Morehouse Lane
Red Hook, NY 12571 USA
Phone: 845-758-0400
Fax: 845-758-2633
Email: curran@proceedings.com
Web: www.proceedings.com

TABLE OF CONTENTS

PREDICTING WATER-SOLUBLE CARBOHYDRATES AND ETHANOL-SOLUBLE CARBOHYDRATES IN COOL-SEASON GRASSES WITH NEAR-INFRARED REFLECTANCE SPECTROSCOPY

K. Prince, B. Goff, L. Lawrence, I. Kagan, and S. R. Smith[1]

Abstract

Grazing animals may require a high or low total nonstructural carbohydrate diet for optimal health and production. Understanding how nonstructural carbohydrates fluctuate in Kentucky pastures and being able to quantify and monitor nonstructural carbohydrates in a timely manner will greatly aid in managing grazing animals in this region. However, using laboratory chemistry to quantify nonstructural carbohydrates can be time consuming and costly. The objective of this study was to develop calibrated equations for faster and easier analysis of water-soluble carbohydrates (WSC) and ethanol-soluble carbohydrates (ESC) in cool-season grasses in Kentucky using near-infrared reflectance spectroscopy (NIRS). Eight cool-season grass cultivars consisting of Kentucky bluegrass, tall fescue, orchardgrass and perennial ryegrass were sampled every 7 to 14 days during the growing season from the University of Kentucky Research Farm in Lexington, KY. Samples were immediately flash frozen after sampling, freeze dried, ground and scanned into FOSS 6500 NIRS autosampler with ISIscan software. WSC and ESC were determined using colorimetric phenol-sulfuric acid assays and served as the basis for NIRS calibration. Based on the statistics of these calibrations, the equations provided robust and accurate estimates of WSC and ESC. This method will aid in simple and timely monitoring of nonstructural carbohydrate levels in pastures in the future, and will allow for the more efficient management of grazing animals.

<u>Introduction</u>: Grazing animals may require a high or low total nonstructural carbohydrate (simple sugars, fructans and starch) diet for optimal health and production. For example, while a diet high in total nonstructural carbohydrates (NSC) may be appropriate for animals with high nutrient needs, horses at risk for laminitis or other metabolic disorders should be fed a diet with limited NSC. Understanding how NSC fluctuate in Kentucky pastures and being able to quantify and monitor these compounds in a timely manner could improve the management of grazing animals in this region. Using laboratory chemistry to quantify NSC of forage pastures can be time consuming and costly. The objective of this study was to develop robust near-infrared reflectance spectroscopy (NIRS) calibrated equations for faster and easier analysis of water-soluble carbohydrates (WSC) and ethanol-soluble carbohydrates (ESC) in cool-season grasses. Since water and ethanol solubilize different nonstructural carbohydrates, having both equations available depending on the type of NSC of interest will be valuable for both research and industry. (Water solubilizes simple sugars and fructan, whereas ethanol solubilizes mainly simple sugars). For example, fructan, specifically, has been linked with the onset of laminitis, so it is of particular interest for horse managers.

---

[1] M.S. Candidate, Department of Plant and Soil Sciences, University of Kentucky; Assistant Professor, Department of Plant and Soil Sciences, University of Kentucky; Professor, Department of Animal Sciences, University of Kentucky; Researcher, United States Department of Agriculture – Agricultural Research Service; Forage Extension Specialist, Department of Plant and Soil Sciences, University of Kentucky.

Nonstructural Carbohydrates and Horses

Laminitis is a debilitating condition that affects the hoof capsule of the horse. Fifty percent of laminitis cases in the United States have been diagnosed as pasture-associated laminitis (PAL) (National Animal Health Monitoring System (NAHMS), 2000), and there is strong evidence linking PAL with excessive intakes of non-structural carbohydrates (Longland and Byrd, 2006), particularly fructan (Van Eps and Pollitt, 2006). Understanding and controlling NSC intake, particularly fructan, in grazing horses is key to not only managing an appropriate nutritional diet, but also in managing and preventing laminitis in at-risk horses.

Near Infrared Reflectance Spectroscopy

Near Infrared Reflectance Spectroscopy (NIRS) has been developed as a time-efficient, valuable tool for forage and feed analysis. NIRS has many uses in agriculture including nutritional quality analysis of forage and feed, pasture management, research and human food industries. In addition to agriculture, NIRS is used in a variety of industries including pharmaceuticals and human medicine (Ciurczak, 2002). Advantages of NIRS when compared to wet chemistry procedures include large sample capacity, minimal preparation of samples, time efficiency, relatively low cost per sample and ease of use.

The NIRS reads the absorbance of each sample in the near infrared region (700-2500 nm) which is then used to develop a unique spectrum (Ruiz, 2001). Calibration equations are developed by combining NIRS spectra with coordinating wet chemistry data (WSC, % DM) of a population representative to the samples that will be analyzed. These two sets of data create a mathematical relationship that can then be used to predict concentrations in unknown samples. Accuracy in calibration development is crucial for accuracy of prediction of unknown samples.

The objective of this study was to develop equations to predict WSC (% DM) and ESC (% DM) to allow for faster and easier analysis of nonstructural carbohydrate concentrations in pastures. Quicker analysis will aid in monitoring and understanding NSC variation and cycles in pastures, ultimately leading to better pasture and animal management.

<u>Materials and Methods</u>: The initial calibration development used cool-season grass samples from a preliminary study in 2014 to begin building the database for WSC and ESC equations. This study sampled two cultivars each of Kentucky bluegrass, orchardgrass, tall fescue and perennial ryegrass from the Forage Variety Trial plots located on the University of Kentucky Research Farm (Lexington, KY). Samples were collected every 7-14 days from mid-April through the end of May. Samples were flash frozen when harvested, freeze dried, ground, and scanned into a FOSS 6500 Near Infrared Reflectance Spectroscopy (NIRS) (FOSS, Inc., Eden Prairie, MN). In 2015, samples were taken from newly seeded research plots. These plots consisted of 10 cultivars representing the same species (orchardgrass and perennial ryegrass: 3 cultivars; Kentucky bluegrass and tall fescue: 2 cultivars) used in 2014 and was arranged as a RCBD. These plots were subdivided to incorporate an N treatment (0 and 115 lbs N/acre) to evaluate carbohydrate variation from soil fertility. Plots were sampled twice daily every two or four weeks from mid-April to early November, and sampling and laboratory methods resembled those of the 2014 study. These samples will be added to the WSC and ESC equations to increase population variation and build robustness of the equations.

Colorimetric, phenol-sulfuric assays were performed to quantify WSC (% DM) and ESC (% DM) (DuBois et al., 1956). All 2014 samples served as the basis for calibration equation development. Equations were developed using modified partial least squares regressions. Regressions were developed and selected based on high R^2 and 1-VR (The 1-VR value for NIRS is the equivalent of an R^2 for a prediction regression) and low standard errors of calibration (SEC) and cross-validation. Equations were cross-validated internally using the leave-one-out method. A math treatment of 2,4,4,1 was selected to optimize regression statistics with the critical T and GH outlier values of 2 and 3, respectively.

Results and Discussion: The WSC (n=63) equation resulted in an estimable range of 1.06% to 22.03% (Table 1). The 1-VR of this equation was 0.90, indicating good predictability of future samples. (With the addition of future sample populations from the 2015 study, the R^2 and 1-VR are likely to increase as the equation becomes more robust.) The ESC equation (n=93) also represents a wide range of estimable ESC values, ranging from 0.49% to 22.79%. The 1-VR for this equation, 0.96, was slightly higher than that of the WSC equation, also indicating good predictability of unknown samples under a wide population range.

To demonstrate the effectiveness of the WSC and ESC equations, the 2014 population was predicted by the equations, and plotted as functions of the actual (wet chemistry) values. These relationships are shown in Figure 1 for WSC and Figure 2 for ESC. The regression equation and R^2 values indicate accuracy of the equation in predicting unknown samples. Both equations have slopes above 90% accuracy, with WSC having a slope of 0.9131 and ESC having a slope of .9585; both regressions also have a R^2 above .90, with WSC having a R^2 of .9125 and ESC having an R^2 of 0.9522.

Based on the equations statistics and regressions of predicted values as functions of the actual values, the WSC and ESC equations indicated strong reliability of accurate predictions of the NIRS equations for this population. Equation statistics typically improve with the addition of more samples, and the 2015 study will add a larger population with wider variability to build robustness of the equations and increase predictability accuracy. Previous NIRS studies of WSC observed better predictability after adding multiple years of samples with wet chemistry to the database used to calibrate equations (Shetty et al., 2012). After a few years of building the equations, the NIRS equations have the potential to become a vital tool in NSC predictions for both research and industry purposes. In both cases, it will streamline NSC analysis by making handling larger sample populations more feasible by minimizing the amount of required wet chemistry analysis, as well as expediting the amount of time it takes to quantify results. Ethanol and water soluble – carbohydrate concentrations in pastures have the potential to change very quickly, therefore having faster turnaround time for sample analysis will improve animal management.

Conclusion: NIRS equations for both WSC and ESC indicated strong predictability for a wide range of values. Both equations have the ability to accurately predict samples within a range of 1% to 20%. Based on available literature and previous studies, typical WSC and ESC for forages fall well within this range (Jensen et al., 2014). In addition to a wide range predictability, both equations had high 1-VR and R^2 values of over .90 and .94, respectively.

In summary, the development of NIRS equations for WSC and ESC will increase the efficiency of quantifying WSC and ESC in cool-season grasses of pastures. Building the equations with additional research samples and using external validation samples from farms would further increase the accuracy and strengthen the reliability of the equations. Having the ability to rapidly and inexpensively monitor WSC and ESC concentrations in cool-season grass pastures will aid in making better informed management decisions for grazing animals.

Literature Cited:

Ciurczak, E. W. 2002. Growth of Near-Infrared Spectroscopy in Pharmaceutical and Medical Sciences. Proc. SPIE 4626 Biomedical Nanotechnology Architectures and Applications.

DuBois, M., K. A. Gilles, J. K. Hamilton, P. A. Rebers, and F. Smith. 1956. Colorimetric Method for Determination of Sugars and Related Substances. Analytical Chemistry. 28(3): 350-356.

Eades, S. C. 2010. Overview of Current Laminitis Research. Veterinary Clinics of North America: Equine Practice. 26(1): 51-63.

Jensen, K. B., P. Harrison, N. J. Chatterton, B. S. Bushman, J. E. Creech. 2014. Seasonal Trends in Nonstructural Carbohydrates in Cool- and Warm-season Grasses. Journal of Crop Science. 54(5): 2328-2340.

Longland, A. C. and B. M. Byrd. 2006. Pasture Nonstructural Carbohydrates and Equine Laminitis. The Journal of Nutrition 136(7): 2099S-2102S.

National Animal Health Monitoring System (NAHMS). 2000. Laminitis in US Horses. Fort Collins CO. USDA.

Ruiz, N. 2001. Near Infrared Spectroscopy: Present and Future Applications (Vol. FT52). N.p.: ASA Technical Bulletin.

Shetty, N., R. Gislum, A.M. Dahl Jensen, and B. Boelt. 2012. Development of NIRS calibration models to assess year-to-year variation in total non-structural carbohydrates in grasses using PLSR. Chemometrics and Intelligent Laboratory Systems. 111 (2012): 34-38.

Van Eps, A. W. and C. C. Pollitt. 2006. Equine Laminitis Induced with Oligofructose. Equine Veterinary Journal. 38: 203–208.

Watts, K. A. and N. J. Chatterton. 2004. A Review of Factors Affecting Carbohydrate Levels in Forage. Journal of Equine Veterinary Science 24(2): 84-86.

Table 1. Equation statistics for quantification of water-soluble carbohydrates (WSC, % DM) and ethanol-soluble carbohydrates (ESC, % DM) from eight cool-season grass cultivars by near-infrared reflectance spectroscopy.

Equation	**n**	**Mean**	**Range**	**SEC**	**RSQ**	**SECV**	**1-VR**
WSC (% DM)	63	11.55	1.06 - 22.03	0.83	0.94	1.09	0.90
ESC (% DM)	93	11.64	0.49 - 22.79	0.41	0.99	0.77	0.96

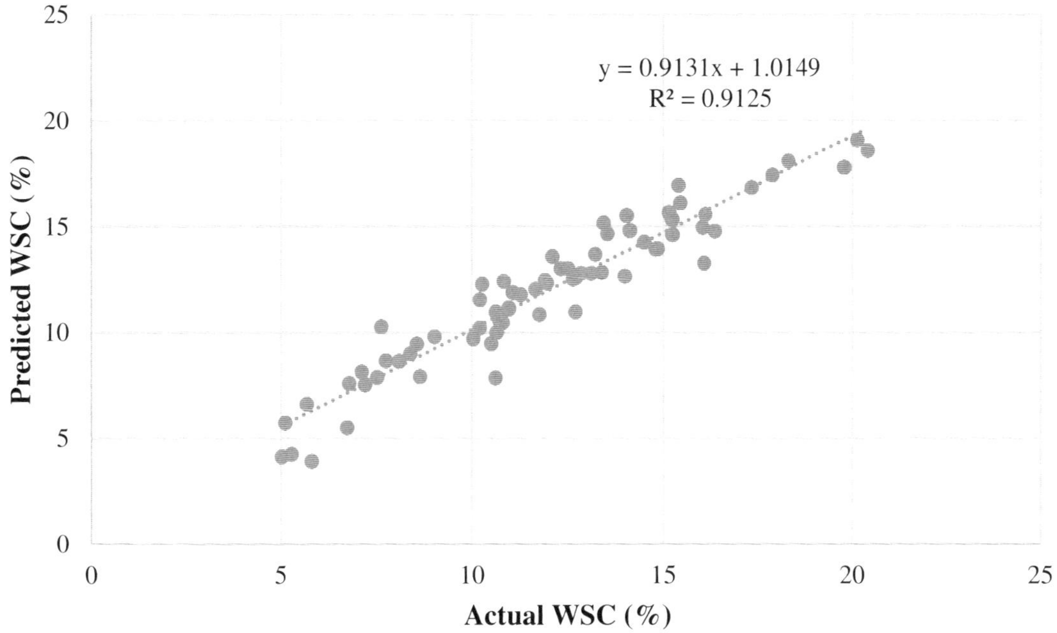

Figure 1. Near infrared-reflectance spectroscopy equation-predicted water-soluble carbohydrates (WSC, % DM) as a function of laboratory (actual) values (WSC) of eight cool-season grass cultivars, spring 2014.

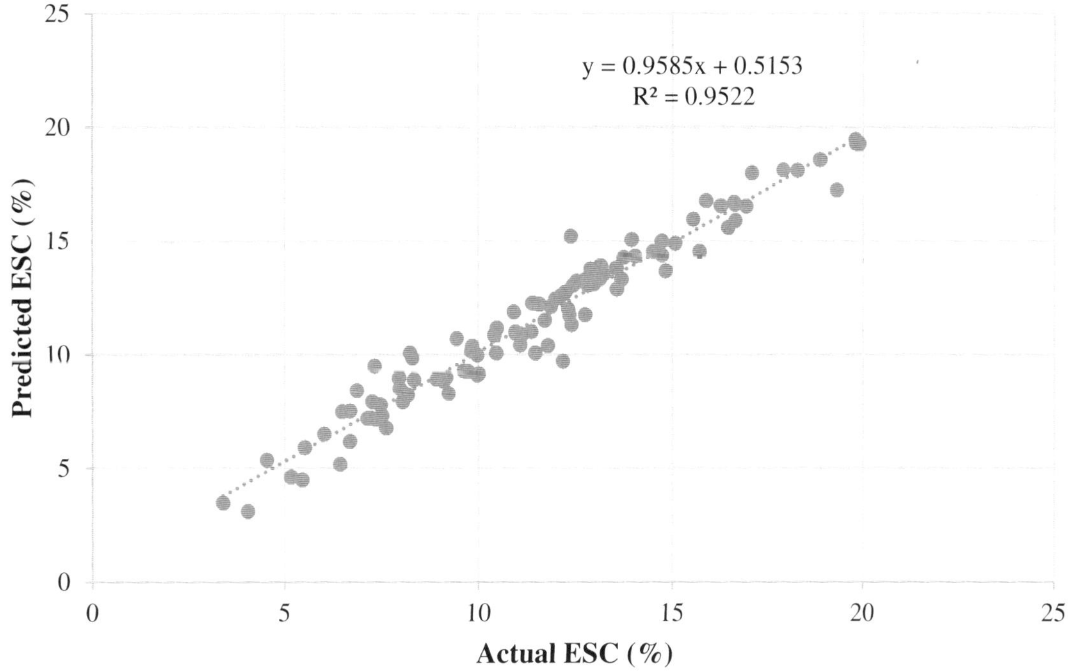

Figure 2. Near-infrared reflectance spectroscopy equation-predicted ethanol-soluble carbohydrates (ESC, % DM) as a function of laboratory (actual) values (ESC) of eight cool-season grass cultivars, spring 2014.

ASSESSMENT OF A BERMUDGRASS CORE COLLECTION FOR COLD AND SHADE TOLERANCE PLUS NITROGEN USE EFFICIENCY

William F Anderson and Freddie Cheek[1]

Bermudagrass (*Cynodon* sp.) is a valuable forage and hay crop in many parts of the world and is now grown as improved pasture on an estimated 2-3 million hectares in southern United States. Bermudagrass hybrids with greater yields, better quality for ruminant digestion, greater disease tolerance, and higher cold tolerance have been accomplished through breeding efforts exploiting the genetic diversity of bermudagrass. Phenotypic data from a large world collection nursery at Tifton, GA was used to form a core collection. The 175 entry core collection was replicated and evaluated in test plots at different locations for a) Cold tolerance in the mountains of North Georgia at Blairsville, GA, (34.84 Lat., -83.93 Long., 590 m Elev.); b) Shade tolerance under 50% shading in Tifton GA (31.48 Lat., -83.53 Long., 108 m Elev.); c) Nitrogen use efficiency in Tifton, GA in 2010 in which each genotype was given either 50 kg/ha equivalent N or 200 kg ha^{-1} equivalent N plus 40 kg/ha P and 100 kg/ha K in each year. For cold tolerance 20 accessions had a higher average yield than Tifton 44 for two years that they survived. For shade tolerance all genotypes declined in yield during the second year of shade. Thicker stem genotypes Florakirk, African Star and breeding lines averaged over 80 g/m total for the second year which is equivalent to 8 Mg/ha. For nitrogen use efficiency total average yield reductions in year one was 35%, and in years two and three, 78% and 73% reductions, respectively. Germplasm was identified for use in breeding programs for these three traits.

[1] William F. Anderson, and Freddie Cheek, Research Geneticist and Research Technician, USDA/ARS, Crop Genetics and Breeding Research Unit, Tifton, GA (bill.anderson@ars.usda.gov)

BREEDING OF NATIVE WARM-SEASON GRASSES TO MAKE THEM AG-FRIENDLY

B.S. Baldwin, J.I. Morrison, J.D. Richwine and J.B. Rushing[1]

Native grasses are rarely used in agriculture, primarily due to seed dormancy, the lack of germination upon planting. While seed dormancy is an advantage in the wild; it is a hindrance to establishment in domestic situations such as pasture renovation, prairie restoration and planting roadside right-of-ways. While there are a number of laboratory methods to ameliorate seed dormancy, the effects are often reversible, or may negatively impact seedling development. This work uses standard breeding methods (phenotypic recurrent selection) to decrease seed dormancy in seven important warm-season grass species native to the southeastern United States: upland and lowland switchgrass, big and little bluestem, purpletop, beaked panicum and eastern gamagrass. Success in meeting our objective has been excellent (0.02 to 94% germination, a 4000 fold increase) in some species and modest (1.0 to 40% and 0.067 to 20% germination, a 40 and 300 fold increase, respectively) in others. A serendipitous side effect of selection was an increase in the speed of seedling emergence (velocity of germination), in addition to an increase in total emergence. Selection reducing seed dormancy allows for quicker emergence and stand establishment, compared to other comparable native grass varieties.

[1] Professor, Dept. of Plant & Soil Science, Mississippi State, MS (bsb2@msstate.edu); Research Assoc. II, Dept. of Plant & Soil Science, Mississippi State, MS (jim46@msstate.edu); Graduate Research Assistant, Dept. of Plant & Soil Science, Mississippi State, MS (jdr211@msstate.edu); Assistant Professor, Coastal Plains R&E Center, Newton, MS (brushing@pss.msstate.edu).

ASSESSMENT OF GAMAGRASS GERMPLASM FROM ACROSS THE SOUTHEASTERN U.S.

B.S. Baldwin, J.I. Morrison, and J.D. Richwine[1]

Gamagrass is a high quality forage with a natural range in the United States from Iowa to Texas. However, the plant is difficult to exploit as a domestic crop. Compared to other grasses it produces relatively few seed per seedhead. Seed ripen sequentially and fall from the plant. Most varieties are composed of seed-clones of a single plant, precluding the ability to make improvements. 'Highlander', a variety widely planted in the Southeast, was selected for relatively high seed number per seedhead and not forage quality. To augment the gamagrass breeding project at Mississippi State University, 172 plants were collected from across six southeastern states. Stands with greater than 20 individuals were targeted for selection and location noted by GPS. Distance between stands was at least 30 miles. Individual proaxes (root crowns) were dug from a single representative plant. All plants were replanted on a three square foot grid for observation and data collection at Starkville, MS. Among characteristics monitored were: number of female flowers per seedhead, onset of flowering, precocious seed germination, gross foliar disease susceptibility, nutritive value, leaf softness and °Brix. Breeding potential will be evaluated in the next six months. Substantial variation existed among individuals for all characteristics measured. Selection of plants with the best combination of characteristics were propagated and placed into a new test for further evaluation.

[1] Professor, Dept. of Plant & Soil Science, Mississippi State, MS (bsb2@msstate.edu); Research Assoc. II, Dept. of Plant & Soil Science, Mississippi State, MS (jim46@msstate.edu); Graduate Research Assistant, Dept. of Plant & Soil Science, Mississippi State, MS (jdr211@msstate.edu).

COMPARION OF TRADITIONAL AND NOVEL NON-DESTRUCTIVE SAMPLING TECHNQIUES FOR SITE-SPECIFIC ASSESSMENT OF BOTANICAL COMPOSITION IN GRASS-LEGUME PASTURES

L.L. Baxter and C.P. West

To better manage forage productivity, it is important to accurately and objectively monitor legume concentrations in grass-legume pastures. Currently researchers rely on subjective visual assessments or labor-intensive botanical hand separations to determine legume content in mixed pastures. Four novel sampling techniques were compared to traditional botanical hand separations to determine their accuracy and reliability in making site-specific assessments of botanical composition in tall wheatgrass-alfalfa [*Thinopyrum ponticum* (Podp.); *Medicago sativa* L.] and Old World bluestem-alfalfa-yellow sweetclover pastures [*Bothriochloa bladhii* (Retz) S.T. Blake; *Melilotus officinalis* L.]. Techniques included visual estimation, photometric (ImageJ and Microsoft® PowerPoint®), and in-field canopy reflectance. Visual and photometric techniques cannot differentiate wheatgrass from alfalfa since their reflected canopy colors are similar in hue and saturation. Visual and PowerPoint analyses were most accurate in determining the legume concentration of Old World bluestem-legume pastures. More thorough residual analyses are required before definitive conclusions can be drawn regarding the accuracy of these procedures. Reflectance data will be analyzed to determine wavelengths correlated with legume and grass components in mixed pastures, with the goal being to apply these findings to automated scanners that offer producers rapid measurements of legume concentration.

Lisa L. Baxter, Department of Plant and Soil Science, Texas Tech University, lisa.baxter@ttu.edu; Chuck P. West, Department of Plant and Soil Science, Texas Tech University, chuck.west@ttu.edu

COMPARISON OF PRODUCTIVITY, EFFICIENCY, AND PROFITABILITY OF GRASS-ONLY AND GRASS-LEGUME BEEF STOCKER GRAZING SYSTEMS IN THE SOUTHERN HIGH PLAINS

L.L. Baxter and C.P. West

The imminent depletion of the Ogallala Aquifer demands innovative alternatives to prevent dramatic losses of income when water levels are insufficient for irrigated row-crop production in the Southern High Plains. With the goal to optimize sustainability and productivity with pasture-based beef stocker systems, a grass-legume (GL) pasture system employing novel forages and grazing techniques was compared to a previously examined grass-only (GO) pasture system. Since protein is often the limiting factor in improving performance of stockers grazing warm-season perennial pastures, higher gains are expected with strategic inclusion of legumes such as alfalfa (*Medicago sativa* L.). Alfalfa was managed as a protein bank so that limited water resources were concentrated on a small area to produce a high-quality supplement for poorer-quality base pastures. Average daily gains between treatments did not differ in year 1. After considering rotation frequency and forage utilization in the first year, a decision was made to increase the duration but decrease the frequency of rotating stockers into the alfalfa. In year two, stocker gains in the GL treatment exceeded those in the GO treatment since longer, more infrequent rotations promoted gains and allowed for better management of high-quality pasture resources. The forage-livestock systems created in this research will serve as alternatives for producers transitioning from irrigated row-crops into more resource-sustaining systems.

Lisa L. Baxter, Department of Plant and Soil Science, Texas Tech University, lisa.baxter@ttu.edu; Chuck P. West, Department of Plant and Soil Science, Texas Tech University, chuck.west@ttu.edu

Interpretive Summary

EXAMINING VEGETATIVE GROWTH OF WELL MANAGED COOL-SEASON FORAGE GRASSES FOR DAIRY CATTLE GRAZING PREFERENCE

Eric D. Billman, Tim D. Phillips[1]

Determining dairy cattle preference of various cool-season forage grasses is vital to understanding future avenues of plant breeding research as well as giving producers immediate benefits, such as identifying what grasses animals most readily consume when given the option to be selective (i.e. low stocking rate and many species/varieties to graze). Maturity differences often bias selection between cool-season grasses, so we eliminated that problem from our study via allowing grasses to become completely vegetative. The objectives of this study were to determine dairy cattle preference of four commonly utilized cool-season forage grass species (orchardgrass, festulolium, perennial ryegrass, and tall fescue) and varieties within these species, as well as determine the cause of any observed preferences or lack of preferences. Our study indicates that under conditions where these cool-season grass species are allowed to reach a completely vegetative state, consistent preference by dairy cattle is not observed. Rather, random and uniform grazing is observed. This is because forage quality differences between various species, while statistically different, are very small, and animals cannot perceive any difference in fiber content, digestibility, or protein. Physical variation such as height may play a larger role, but this was not observed in our trials.

[1]Graduate Student, Department of Plant and Soil Sciences, University of Kentucky; and Associate Professor of Forage Grass Breeding, Department of Plant and Soil Sciences, University of Kentucky.

EXAMINING VEGETATIVE GROWTH OF WELL MANAGED COOL-SEASON FORAGE GRASSES FOR DAIRY CATTLE GRAZING PREFERENCE

Eric D. Billman, Timothy D. Phillips[1]

Abstract

Identification of forage pasture grasses that are more readily consumed than others during grazing is vital to determining traits that can be of interest to plant breeders to improve animal intake and digestibility. Producers also can benefit from such studies to identify what species, and varieties within a species, are more readily consumed and thus will increase their animal production levels. The objective of this study was to determine dairy cattle preference amongst four species of cool-season forage grasses: orchardgrass (*Dactylis glomerata* L.) (8 varieties), tall fescue [*Lolium arundinaceum* (Schreb.) Dumort.] (5 varieties), perennial ryegrass (*Lolium perenne* L.) (5 varieties), and festulolium [x *Festulolium braunii* (K. Richt.) A. Camus.] (6 varieties); to determine the most palatable variety and species to dairy cattle. After six grazing trials, consistent results with regard to animal preference were not found. However, forage quality analysis shows that the entries (varieties and species), while statistically different ($P<0.1$), did not vary greatly in the measured constituents. This is likely due to the elimination of confounding maturity differences, and plots being maintained and managed effectively and thus explains why the animals did not show consistent preferences of one species or variety over another.

Introduction: Cool-season forage grasses are widely used in dairy systems as a major source of forage biomass. Of particular interest in determining preference are four species that are most commonly found growing in dairy pastures in the central and northern United States: orchardgrass (*Dactylis glomerata* L.), tall fescue [*Lolium arundinaceum* (Schreb.) Dumort.], perennial ryegrass (*Lolium perenne* L.), and festulolium [x *Festulolium braunii* (K. Richt.) A. Camus.]. Each species has unique morphological traits and chemical compositions that could affect dairy cattle preference, either favorably or adversely. For example, orchardgrass has a rapid rate of regrowth following mowing or grazing, and often has a greater sward height than other species, which has been shown to be correlated to animal preference (Smit et al., 2006). On the other hand, tall fescue can have rigid and coarse leaves that might be undesirable to animals when selecting which plants to graze first. Within species, cultivar differences are less pronounced than among species differences, but do still exist. Previous work has shown that within species variation of chemical composition in perennial ryegrass cultivars led to significant preference of cultivars exhibiting higher water soluble carbohydrates and lower NDF values (Smit et al., 2006), as well as tetraploid perennial ryegrasses being more preferred than diploids due to pregrazing biomass availability (Solomon et al. 2014; Hageman et al., 1993; Vipond et al., 1993).

---

[1]Graduate Student, Department of Plant and Soil Sciences, University of Kentucky; and Associate Professor of Forage Grass Breeding, Department of Plant and Soil Sciences, University of Kentucky.

Similar preference studies have been conducted, but are usually done only on a within cultivar basis on a single species (Waldron et al. 2010), or across a wide range of species and plant families, i.e. grasses and legumes (Horadagoda et al., 2009). Limited knowledge exists on preference within the cool-season grasses exclusively across multiple species, and even less on how dairy cattle preferences change across the growing season, based on morphological changes among different cool-season species.

Our primary research question is: do dairy cattle exhibit preferences among cool-season forage grasses, and if so, what are the causes of these preferences? In response to this, we have two major objectives: (i) determine dairy cattle preference of cool-season forage grass species and varieties; (ii) compare phenotypic traits among entries to determine the causes of these preferences.

Materials and Methods: Four cool-season forage grass species, orchardgrass, tall fescue, perennial ryegrass, and festulolium, served as the four species entries for this study. Within these, varieties were randomly selected from available germplasm to serve as individual entries, each nested within a particular species. Twenty four total varieties and cultivars were used, however the number of varieties per species was unbalanced. Eight commercial varieties of orchardgrass were used: Persist, Potomac, Profit, Harvestar, Benchmark Plus, Prairie, Tekapo, and Prodigy. A total of six varieties were used to represent festuloliums: Perun, Duo, Barfest, SpringGreen, KYFA1015 and KYFA1016 from the University of Kentucky. A total of five tall fescue cultivars were used: Barianne, Barolex, Jesup EF, Select, and the cultivar KYFA9611 from the University of Kentucky. Finally, five varieties of perennial ryegrass were chosen: Linn, Granddaddy, BG34, Power, and Calibra.

Using these 24 entries, a randomized complete block design containing six replications consisting of 144 total plots was planted at the University of Kentucky's Spindletop Research Farm in Lexington, KY in September of 2013. Each plot was 3x10 feet long, with border alleys intersecting each range at the ends of the plots. The total dimensions of the study are 80ft long by 54 feet wide. Plots were planted with a five-row Carter self-propelled, walk-behind planter at a seeding rate of 20 lbs/acre in order to create sward plots for each entry.

An important goal of this study was to eliminate potential bias of animal preference due to varying maturity rates (timing of reproductive growth) of the different species and within-species cultivars of cool-season grasses used in this study. The method for controlling this bias is adapted from a similar study where animals are only grazing on vegetative regrowth, rather than the initial flush of reproductive growth in early spring (Hitchcock et al., 1990). In each year of the study all plots were allowed to reach 50-75% seed-head emergence from the boot stage. At this stage of growth, the entire study was cut at a height of three inches using a Hege 212 forage harvester to remove all of the reproductive growth and allow for only vegetative regrowth to occur and be grazed the remainder of the growing season. Accommodating for this aspect of the study meant waiting until the third week of May in both 2014 and 2015 in order to cut the crop and return all of the plots to strictly vegetative growth.

In both 2014 and 2015 the dairy herd consisted of yearling Holstein-Friesian heifers. Each session lasted for a period of 5-6 hours, using 4 animals each time, depending on how rapidly the animals consumed the available forage. We aimed to achieve approximately 50% consumption of available forage.e

Of most importance in determining preference were Pre-Grazing and Post-Grazing dry matter yields of each plot. These were taken before and after each grazing session using Ariens 22 inch deck-width push mowers, and mowing across the 3ft. width of each plot. Each sample was then bagged and weighed for fresh and dry weight to determine percent dry matter. The difference between the pre-grazing and post-grazing dry weights was used to determine the percentage of total plot consumption. For our purposes, the greater the percentage consumed, the greater the preference for that plot. Pre-grazing and sward heights were also taken for each plot. Three measurements per plot were taken for each grazing session and averaged to get a mean height for each plot.

After weighing of samples for yield data, each was ground to prepare for forage quality analysis. Samples were first ran through a Wiley Mill grinder using a 4mm mesh screen, followed by a second grinding using a UDI Cyclone Mill with a 1mm mesh screen. These samples were then placed into cups and analyzed for forage quality components using a FOSS 6500 NIRS (Near Infrared Spectroscopy) machine. Samples were scanned into the FOSS 6500 to obtain spectrum data, which was then used to determine samples to run for wet lab chemistry to develop a calibration for NDF (neutral detergent fiber), ADF (acid detergent fiber), IVTD (in-vitro true digestibility, and CP (crude protein).

Results and Discussion: Over the course of six grazing trials over two years, consumption was only significant ($P<0.10$) at three out of six trials. Consistent preference was not observed, as the most preferred species changed and no pattern was found (Figure 1). Significant differences in yield, and dry matter content were also not consistent among trials (data not shown). Height was consistently significant ($P<0.10$) and orchardgrass was always the tallest species (data not shown). Forage quality values for NDF, ADF, IVTD, and CP while significantly different, show a very narrow range in mean values between species (Figure 2 A, B, C, and D respectively).

Forage quality analysis lends some insight into why we do not see consistent preference differences between species or varieties within species. As with morphological variables, correlations to forage quality variables were not biologically relevant, despite results to the contrary in previous research (Smit et al., 2006). Forage quality values for NDF, ADF, IVTD, and CP do show statistically significant differences between species at most trials. However, across all trials, the maximum difference in NDF value between species is 5%. ADF, IVTD, and CP differences are even smaller, maxing out at about 3% between species at any one trial for ADF, 2% for IVTD, and 2% for CP. While statistically significant, these differences are likely not biologically significant, in that the animals cannot differentiate between forage that is so similar in their forage quality components. Animal performance differences may be observed with these small differences in quality, but the animal's willingness to consume forages likely will not change. Varieties within each species show similarly small differences between means, with ranges of for quality constituents never exceeding 5% at any one trial (data not shown).

These small relative differences in forage quality components between species and varieties at various trials raise the question of why the variation between entries is small. One would assume that with 24 entries consisting of a wide range of genetic material and differing morphologies that species would be highly significant for preference at each trial. As mentioned in our methodology, one of the primary factors we eliminated from this study was maturity differences. The differing growth rates of different cool-season forage grass species is well documented, and varieties selected and bred for differing maturity rates is common. By allowing our plots to begin their reproductive cycles prior to any grazing, and subsequently removing that growth, we effectively put all of the entries on an even playing field in terms of their growth habit. At the time of the first grazing in both 2014 and 2015, almost all of the plots were entirely vegetative, with little to no stem material being present. This inherently results in higher quality forage, as seen by the NDF, ADF, IVTD, and CP values at each trial, which are good values for cool-season forage grasses. It has been observed that forage quality values for NDF and IVTD decline in acceptability as plants mature (Karn et al., 2006). Because we eliminated maturity differences, this could explain the lack of large difference in measured variables.

In conclusion, we have shown that harvesting forage in its vegetative state leads to a high quality forage that is relatively uniform in quality, even across multiple cultivars and species. Adding good weed control and adequate fertilization to this can also aid in vastly improving overall forage quality, and serves to make the difference between different cool-season grasses smaller. Producers wishing to improve the intake of their cattle on previously established pasture can implement these strategies in order to allow for more efficient utilization of their pasture. Plant breeders should recognize that the differences in forage quality between cool-season grasses in vegetative state of growth are small, and that they should focus on selecting for improved morphological traits such as biomass and height if working with species in a vegetative state.

Literature Cited:

Hitchcock, R. A. M., R. B.; Bradley, N. W.; Wahab, A. A.; Dougherty, C. T. . (1990). Forage composition and intake by steers grazing vegetative regrowth in low endophyte tall fescue pasture. *Journal of Animal Science, 68*, 2848-2851.

Horadagoda, A. F., W J; Nandra, K S; Barchia, I M. (2009). Grazing preferences by dairy cows for 14 forage species. *Animal Production Science, 49*, 586-594.

Karn, J. F., Berdahl, J. D., & Frank, A. B. (2006). Nutritive Quality of Four Perennial Grasses as Affected by Species, Cultivar, Maturity, and Plant Tissue. *Agronomy Journal, 98*(6), 1400-1409.

Smit, H. J., Tamminga, S., & Elgersma, A. (2006). Dairy Cattle Grazing Preference among Six Cultivars of Perennial Ryegrass. *Agronomy Journal, 98*(5), 1213-1220.

Solomon, J. K. Q., B. Macoon, D.J. Lang, R.C. Vann, and S. Ward. (2014). Cattle grazing preference among tetraploid and diploid annual ryegrass cultivars. *Crop Science, 54*, 430-438.

Vipond, J. E., Swift, G., McClelland, T.H., Fitzsimmons, J., Milne, J.A. and Hunter, E.A. (1993). A comparison of diploid and tetraploid perennial ryegrass and tetraploid ryegrass/white clover swards under continuous ship stocking at controlled sward heights. *Animal Production, 48*, 290-300.

Waldron, B. L., Davenport, B. W., Malechek, J. C., & Jensen, K. B. (2010). Relative Cattle Preference of 24 Forage Kochia () Entries and Its Relation to Forage Nutritive Value and Morphological Characteristics. *Crop Science, 50*(5), 2112-2123.

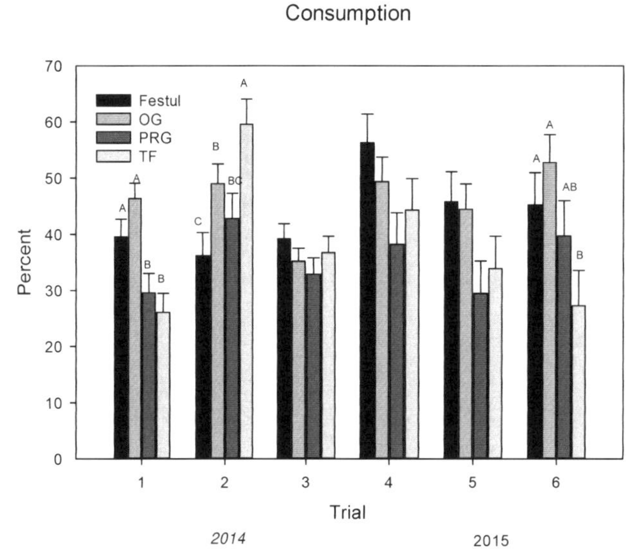

Figure 1. Percent consumption of festulolium (Festul), orchardgrass (OG), perennial ryegrass (PRG), and tall fescue (TF) at six different grazing trials. Trials 1-3 occurred in June, July, and Septmeber of 2014. Trials 4-6 Occurred in June, July, and September of 2015.

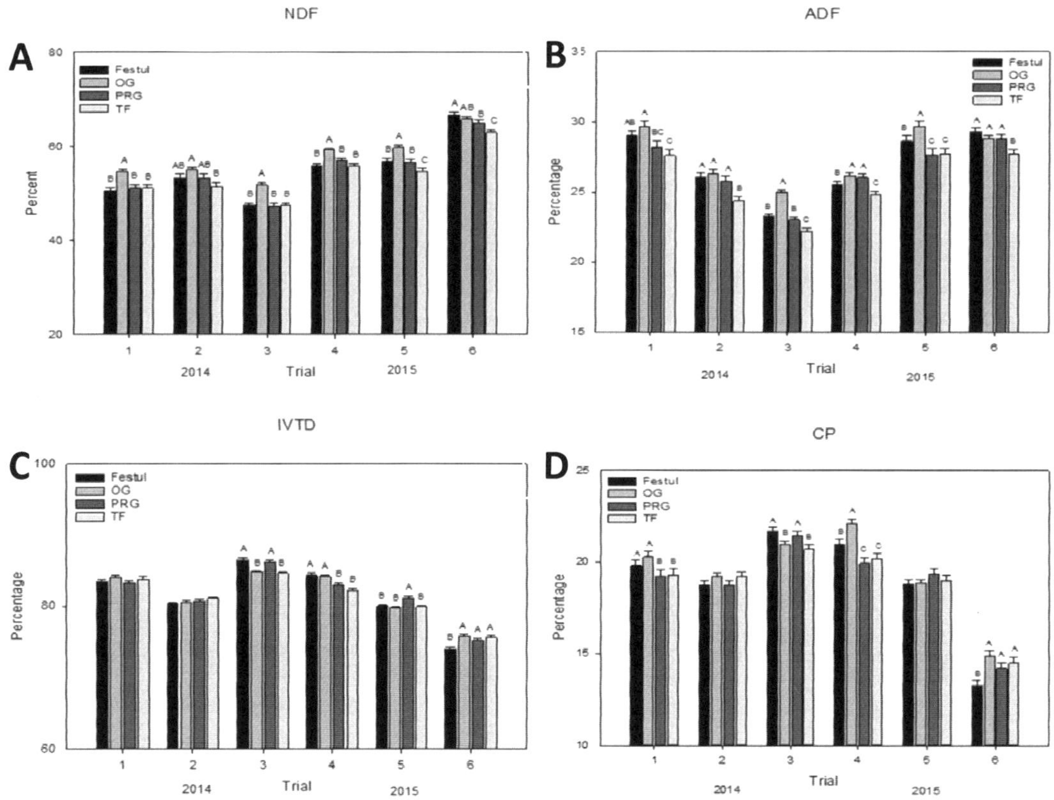

Figure 2. A) NDF content of festulolium (Festul), orchardgrass (OG), perennial ryegrass (PRG), and tall fescue (TF) at six different grazing trials. B) ADF content of festulolium (Festul), orchardgrass (OG), perennial ryegrass (PRG), and tall fescue (TF) at six different grazing trials. C) IVTD content of festulolium (Festul), orchardgrass (OG), perennial ryegrass (PRG), and tall fescue (TF) at six different grazing trials. D) CP content of festulolium (Festul), orchardgrass (OG), perennial ryegrass (PRG), and tall fescue (TF) at six different grazing trials. Trials 1-3 occurred in June, July, and Septmeber of 2014. Trials 4-6 Occurred in June, July, and September of 2015.

PERFORMANCE OF DAIRY HEIFERS REARED ON PASTURE VS. CONFINEMENT

G.E. Brink[1], W.K. Coblentz[2], and P. Hoffman[3]

Rearing dairy heifers on pasture may have animal performance advantages over confinement rearing. We compared growth and first-lactation milk production of Holstein dairy heifers reared on pasture or in confinement. Thirty-six weaned calves (368 lb mean BW; body weight) were divided into six groups of equal total weight and randomly assigned to rotationally graze vegetative-stage, cool-season, grass-legume paddocks for two grazing seasons or be fed a TMR (total mixed ration) in confinement pens (three replicates of each). Pasture heifers were also fed a commercial calf-starter mix during the grazing season (0.5% BW/day). Pasture DM (dry matter) intake was estimated as pre-graze – post-graze DM. Confinement DM intake was measured as pre-fed – orts DM. All heifers were fed the same TMR during the winter. Milk production of cohorts was measured over a 205-day lactation and intake was measured at 50, 100, 150, and 200 days-in-milk to calculate feed efficiency. Heifers reared on pastures gained weight more slowly than those in confinement during both grazing seasons (1.4 vs. 1.7 lb/d), but weight of all heifers was similar at freshening. Heifers reared on pasture produced more milk than those in confinement (18,200 vs. 17,000 lb), had lower milk somatic cell count (164,000 vs. 200,000), and similar feed efficiency during lactation.

[1] Research Agronomist, USDA-ARS, U.S. Dairy Forage Research Center, Madison, WI 53706; [2] Research Dairy Scientist, USDA-ARS, Institute for Environmentally Integrated Dairy Management, Marshfield, WI 54449; [3] Dairy Technical Specialist, Vita-Plus, Inc., Madison, WI. gebrink@wisc.edu

A SURVEY OF GRASS-FINISHED BEEF PRODUCERS IN THE NORTHEASTERN U.S.

Jasmine Dillon and Alan Rotz

Sixty-nine self-identified grass-finished beef producers responded to an online survey implemented via email listservs in the Northeast. Fifty-four records from producers in Maine, Massachusetts, New Hampshire, New York, Pennsylvania, Vermont, and West Virginia were included in the analysis after excluding incomplete entries and entries from producers who were not selling grass-finished beef or cattle destined for the grass-finished beef market. The average farm managed 25 cows and 18 finishing animals on 113 acres. The average grass-finished animal was finished and marketed for beef at 23 months of age and 1,021 lb. Seventy-three percent of the total land represented by the survey was grazed and 27% was cropped. Of the producers producing hay from pasture on the farm, 57% reported replanting at some interval and 71% reported using no-till management practices. Eighty-seven percent of farms raised calves from birth to finish, while 13% of farms were finishing operations only. Fifty-two percent of producers reported purchasing feed for the herd, and dry hay was the most commonly purchased feed. Beef products were most commonly marketed directly to the consumer (94% of producers), although 23% of producers marketed their beef through multiple outlets. Sixty-three percent of producers reported marketing additional species from the farm. Production information provides a basis to evaluate and improve the sustainability of grass-finished beef.

Jasmine Dillon is a PhD Candidate in the Department of Animal Science at The Pennsylvania State University, University Park, PA (jad520@psu.edu). Alan Rotz is an Agricultural Engineer at the USDA-ARS Pasture Systems and Watershed Management Research Unit, University Park, PA (al.rotz@ars.usda.gov)

MANAGING NATIVE FORAGES FOR DOMESTIC LIVESTOCK GRAZING AND GRASSLAND BIRDS HABITAT IN SOUTH CAROLINA

Jill Epley

The United States Forest Service, Clemson Extension, Duke Energy, South Carolina Department of Natural Resources, National Wild Turkey Federation and the USDA-Natural Resources Conservation Service have partnered together to provide funding, technical support, and management to restore approximately four hundred acres of introduced pasture back to native grasses and forbs in Union County, South Carolina called the Delta Project. The purpose is to provide outreach and education about native ecosystems, and to discuss and demonstrate methods of managing grazing of domestic livestock while creating and maintaining habitat for Grasshopper Sparrows, Bobwhite Quail, Bobolinks and many more grassland and migrating birds. The partners listed above will provide hands on outreach and education during the site preparation and establishment of native grasses, prescribed burning, prescribed grazing, and wildlife habitat management on the site for farmers, ranchers, and the general public.

Conservation Management Plan Timeline: 2015- Chemical site preparation of Tall Fescue.

2016- Prescribed burn and chemical site prep of Tall Fescue and Johnson grass. 2017-Native grass seed planting. 2018-Livestock and wildlife watering facility establishment and post planting grass stand management. 2019-Introduction of Livestock and Prescribed grazing management for grassland birds.

Resource Conservationist, USDA-NRCS, Columbia, SC (jill.epley@sc.usda.gov)

USING ASH CORRECTED aNDFom VERSUS aNDF FOR GRASS AND LEGUME SILAGE FIBER ANALYSIS

S.A. Flis and P. K. Sirois[1]

In 2015 there were changes to reporting and analysis of neutral detergent fiber (NDF). On forage analysis reports NDF was changed to aNDF and the option for having ash corrected aNDFom was added to analysis services. In the 1990's amylase, to breakdown starch, and sodium sulfite, to remove protein, were added to the neutral detergent solution to clean up the residue and give a more accurate representation of fiber in the sample. In 2015, to better represent the analysis coming from a lab, the "a" was added to NDF to indicate that amylase is used in the procedure. Performing aNDF analysis on an "organic matter" or "ash free" basis (aNDFom) corrects fiber concentration for ash contamination. Laboratories and consultants have observed increasing ash concentrations in forage samples. Ash, % DM, in 2001, was 10.7 ± 1.9 and 8.5 ± 2.2 for legume and grass silage, respectively. In 2014, average ash, % DM increased to 11.9 ± 2.3 and 9.9 ± 2.9 for legume and grass silage, respectively. Between January and July 2015 the Dairy One Forage Lab analyzed 2,108 legume silage and 3,151grass silage samples for aNDF and aNDFom. Legume silage averaged 45.3 aNDF, % DM and 43.3 aNDFom, % DM. Grass silage averaged 59.5 aNDF, % DM and 56.7 aNDFom, % DM. This is a difference of 1.98 % for legume silages and 2.77 % for grass silages. Elevated ash values can contribute to elevated aNDF values leading to underestimation and underfeeding of fiber. Feeding low fiber diets can result in animal health problems like ruminal acidosis, foot ulcers, and milk fat depression.

[1]Feed and Crops Support Specialist, Dairy One, Ithaca, NY (sally.flis@dairyone.com); Laboratory Manager, Dairy One, Ithaca, NY (paul.sirois@dairyone.com).

AGRONOMIC AND ECONOMIC IMPLICATIONS OF SEEDHEAD SUPPRESSION DOWN ON THE FARM

E. S. Flynn, P. L. Burch, and Glen Aiken[1]

Introduction: Tall Fescue (Lolium arundinaceum (Schreb.)) is cool-season perennial grass found in pastures throughout the Eastern US, most commonly in the transition zone. It's tolerance to extreme temperatures, drought, poor soil fertility, heavy grazing, and pests have made it a reliable forage base for livestock pastures. Released in 1943, 'KY-31' tall fescue was quickly adopted by cattle producers and now inhabits an estimated 40 million acres in the US.

Soon after its adoption cattle producers began to notice a reduction in animal performance and signs that animals were heat stressed following tall fescue consumption. These symptoms were an indication of an animal disorder that would later be called "fescue toxicosis". In the late 1970s it was discovered that the issues with tall fescue were actually the result of a fungus (Neotyphodium coenophialum) living within some plants, which produced a toxic class of compounds termed ergot alkaloids. We refer to these tall fescue plants that contain this fungus as being endophyte infected (E+) and those that do not as being endophyte free (E-).

Fescue toxicosis is estimated to cost the livestock industry 1$ billion annually. Economic losses are largely associated with reductions in pregnancy rates, milk production, and average daily gains in stocker cattle and poor calf weaning weights (Table 1). Other losses include lower sale prices and higher feedlot receiving costs related to cattle appearance and health.

Seedhead Suppression: Chaparral™ is a broad spectrum herbicide approved for control of susceptible broadleaf weeds and woody plants on rangeland and permanent grass pastures. It contains aminopyralid (2-pyridine carboxylic acid, 4-amino-3,6-dichloro-) and metsulfuron-methyl {methyl 2-[[[[(4-methoxy-6-methyl-1,3,5- triazin-2-yl)-amino]carbonyl] amino]sulfonyl]benzoate}. Timely spring applications of Chaparral on tall fescue pastures inhibit grass maturation, keeping the plants in a vegetative state for the remainder of the growing season.

Both day length and temperature can affect seedhead emergence in tall fescue. Day length of 12 hours or more triggers seedhead formation and temperature influences growth and development. It is best to apply Chaparral starting 3 weeks before seedhead emergence to when tall fescue is in the boot stage. Chaparral application to tall fescue in the US should begin around April 10th and end at the boot stage but could begin as early as March 20th in the south when 12 hour or more day length exists and tall fescue growth has initiated.

---

[1]Field Scientists, Crop Protection Research and Development, Dow AgroSciences LLC, Lees Summit, MO 64082 and Christiansburg, VA 24073; and Research Animal Scientist, Forage Animal Production Research, USDA-ARS, Lexington, KY 50546.

For optimal seedhead suppression apply 2 oz/acre of Chaparral with a 0.25% volume per volume, high quality non-ionic surfactant (NIS). Do not tank-mix with other herbicides as this may reduce the level of seedhead suppression.

The Benefits of Seedhead Suppression: When seedheads are suppressed and weeds are controlled with Chaparral an increase in grass species diversity is usually observed. This increase in diversity along with the better nutritive value of tall fescue, and reduced consumption of ergot alkaloids has led to higher pregnancy rates in cows, higher 205-day adjusted weaning weights in calves (Table 2), and improved average daily gains in steers (Table 3). Tall fescue pastures where seedheads have been suppressed by Chaparral have higher crude protein, *in vitro* dry matter digestibility (Table 3), and water soluble carbohydrates than pastures not treated with Chaparral. In addition, herbage dry matter intake by cattle is greater and cattle hair is shorter and smoother compared to cattle feeding on tall fescue not treated with Chaparral. (Turner et al. 1990a, 1990b; Aiken et al., 2012).

Where does seedhead suppression with Chaparral fit in your operation?: Chaparral is an herbicide that suppresses tall fescue seedhead emergence. Chaparral for seedhead suppression works in rotational grazing systems and other grazing systems were quality forage is needed in late spring – early summer. Seedhead suppressed tall fescue maintains higher forage quality in spring and early summer when compared to unsuppressed tall fescue. This can allow cattle producers to keep up with spring grass production and maintain availability of high quality forage for a longer period during the summer. It also gives producers the ability to rotate to higher quality fescue pastures during a time of year when tall fescue is usually more mature and has low forage quality.

For tall fescue pastures where Chaparral is being used to suppress seedheads it is advised that no more than 50% of total acres be treated in a single season or that applications not be made on the same acres over two sequential years. Chaparral will cause some leaf yellowing in tall fescue for a period of 7 to 14 days with normal growth resuming in 3-4 weeks; however, it is not advised that Chaparral be used on pastures for seedhead suppression with poor soil fertility as the length and severity of yellowing will be prolonged.

Seedhead suppressed pastures produce less total herbage mass early in the season due to the lack of a seedheads. However, leaf dry matter can actually yield as much or more as unsuppressed pastures since older leaves on fescue plants begin to senesce after seedhead production. Herbage mass in suppressed pastures may again decrease in late summer due to increased forage utilization by less stressed cows and their calves (Boyer et al, 2015). It is recommended that producers rotationally graze pastures to prevent overgrazing and implement a spring and fall soil fertility program to maintain pastures with high quality grass forage. Remember, reducing the stress of fescue toxicosis in cattle will increase their dry matter intake and put more pressure on your pastures.

Literature Cited:

Aiken , G.E, B.M. Goff, W.W. Witt, I.A. Kagan, B.B. Sleugh, P.B. Burch and F.N. Schrick. 2012. Steer and plant responses to chemical suppression of seedhead emergence in toxic endophyte-infected tall fescue. Crop Sci. 52:960-969.

Boyer, W.F., G.W. Webb, and E.S. Flynn. 2015 Cow-calf response to seedhead suppressed tall fescue pastures in Southern Missouri. Master Thesis (2015).

Goff, B.M., G.E. Aiken, W.W. Witt, P.B. Burch, E.S. Flynn. 2012a. Effects of seedhead suppression of endophyte-free tall fescue in grass mixtures on steer performance and nutritive values. PhD Dissertation (2012).

Goff, B.M., G.E. Aiken, W.W. Witt, B.B. Sleugh, P. Burch,. 2012b. Steer consumption and Ergovaline recovery from in vitro digested residues of tall fescue seedheads. Crop Sci. 52:1437-1440

Paterson, J., C. Forcherio, B. Larson, M. Samford, and M. Kerley. 1995. The effects offescue toxicosis on beef-cattle productivity. J. Anim. Sci. 73:889-898.

Turner, K.E., J.A. Paterson, J.S. Kerley, and J.R. Forwood. 1990a. Mefluidide treatment of tall fescue pastures: Intake and animal performance. J. Anim. Sci. 68:3399–3405.

Turner, K.E., J.A. Paterson, J.S. Kerley, and J.R. Forwood. 1990b. Mefluidide treatment of tall fescue pastures: Forage quality. J. Anim. Sci. 68:3406–3411.

Table 1. [a]Summary of the effect of E+ tall fescue on cattle. Data derived from multiple research trials were pastures contained 70% or more E+ tall fescue

Performance Metric	**Effect on Production**	
Pregnancy rates	Decreased	15-40%
Milk production	Decreased	25%
Weaning weights	Decreased	65- 85lbs
Time spent grazing	Decreased	20%
Forage intake	Decreased	25-40%
Average daily gain	Decreased	0.3-1.2 lbs/day
Water usage	Increased	25%
Body temperature	Increased	1-4°F

[a]Paterson et.al, 1995

Table 2. Effect of seedhead suppression on 205 day adjusted weaning weights of calves and breed back of cows. Data were collected from three different locations owned by Whitesell Land and Cattle Co. At each location the pastures were split into 2 paddocks so that half the herd would be on a Chaparral-treated paddock (tall fescue seedheads suppressed) and the other half would be on a GrazonNext HL treated paddock (non-suppressed).

Location	205 Day Adjusted Weaning Wt.			Cow Pregnancy Rates	
	Chaparral Treated	Non-suppressed	difference	Chaparral Treated	Non-suppressed
Farm 1	473 lbs	418 lbs	+55lbs	95%	80%
Farm 2	483 lbs	463 lbs	+20lbs	95%	70%
[a]Farm 3	476 lbs	459 lbs	+17lbs	Equal at 91%	

[a]*Heavy spring grazing on treated and untreated paddocks resulted in seedhead suppression through grazing, thus the non-treated paddock behaved similarly to the suppressed paddock. (Adapted from Boyer et al., 2015)*

Table 3. Effect of seedhead suppression on stocker cattle and forage quality of pastures grazed. Data were collected from 2009-2012 in three separate studies conducted by the USDA-ARS (Lexington, KY)

	Chaparral Treated	Non-suppressed
2009-2010		
Steer ADG (lbs/steer/day)	2.1 lbs	1.48 lbs
Crude protein	14.4%	11.6%
In vitro dry matter digestibility	78.6%	71.7%
2011-2012		
Steer ADG (lbs/steer/day)	2.1 lbs	1.74 lbs
Crude protein	14.2%	9.9%
In vitro dry matter digestibility	72.2%	66.4%
2011-2012		
Steer ADG (lbs/steer/day)	2.29	1.96
Crude protein	18.6	14.8
In vitro dry matter digestibility	76.1	72.6%

(Aiken et al., 2012; Goff et al., 2012a; Goff et al., 2012b)

ISOFLAVONE RUMEN METABOLITES; A MISSING LINK IN THE BENEFITS OF LEGUMES ON GRAZING ANIMAL PRODUCTION

M.D. Flythe, G.E. Aiken, I.A. Kagan[1]

Clovers are widely used to add protein to ruminant diets. Legumes also produce a class of small molecules called isoflavones which have estrogenic properties and can interfere with reproduction, but they also have benefits. Potential benefits of clover isoflavones include improvements in rumen function and relaxation of blood vessels. In laboratory tests, red clover (*Trifolium pratense*) extracts decreased amino acid degradation by rumen bacteria. The active component in the extracts was an isoflavone, biochanin A, which inhibited the same group of amino acid-degrading bacteria as ionophore antibiotics. The results of feeding trials supported the idea that biochanin A could increase weight gain, as do growth-promoting antibiotics. The spring and fall trials involved grazing steers that were or were not supplemented with dried distillers grains with and without biochanin A (mixed with the grains to a level commensurate with a 33% red clover diet). Average daily gain was greater in the groups that received biochanin A. Additionally, rumen bacteria metabolize isoflavones, and some of those metabolites are known to dilate blood vessels. Vasodilation would be beneficial because many ruminants are exposed to ergot alkaloids that cause chronic vasoconstriction (*i.e.* fescue toxicosis). Goats were administered toxic tall fescue and red clover extracts, and two arteries in each goat were observed with ultrasonography. The clover extract decreased vasoconstriction caused by the tall fescue extract. These results may explain many of the benefits seen in animals grazing clovers.

[1]Research Microbiologist, USDA-ARS, Forage-Animal Production Research Unit (michael.flythe@ars.usda.gov); Supervisory Animal Scientist, USDA-ARS, Forage-Animal Production Research Unit (glen.aiken@ars.usda.gov); Research Plant Physiologist, USDA-ARS, Forage-Animal Production Research Unit (isabelle.kagan@ars.usda.gov).

YIELD, PERSISTENCE, AND NUTRIENT HARVEST OF TWO CULTIVARS AND TWO BLENDS OF SEEDED BERMUDAGRASS IN A SWINE EFFLUENT SPRAY FIELD

B.L. Spearman, S.R. Freeman, and M.H. Poore

To determine yield potential, persistence, and potential nutrient removal of seeded bermudagrass cultivars (Cheyenne and Mohawk) and two seeded blends (Laredo and Ranchero Frio), four plots each were planted in a swine effluent spray field in Tar Heel, NC. Measurements were taken over four growing seasons, including during establishment, to estimate dry matter (**DM**), nutrient concentration, and weed contribution to DM. Yields of 10,885 and 10,751 lb DM/A for Cheyenne and Ranchero Frio exceeded those for Laredo and Mohawk (10,089 and 10,080 lb/A, respectively). Yield decreased after the second year. Weed contribution to DM was large during establishment, negligible the second year, and then grew greater (18.7, 0.3, 12.8, and 21.3% for the first to fourth seasons, respectively). Weed DM was greater in Cheyenne than Mohawk or the blends (17.2, 11.7, 12.2, and 12.0% for Cheyenne, Laredo, Mohawk, and Ranchero Frio, respectively). Nitrogen and phosphorus yields were proportional to DM. Nitrate concentration was greater in forage from Cheyenne and Ranchero Frio than Laredo and Mohawk (0.58, 0.60, 0.43, and 0.47% of DM, respectively). Lower yields after the second growing season resulted in average yields closer to the realistic yield estimate for common bermudagrass than hybrid bermudagrass. Ranchero Frio and Cheyenne were better suited to this site based on their greater DM yield across years. Ranchero Frio, with less weed DM the final year, may be best at this site.

Becky L. Spearman, Coop. Extension Service, NC State Univ.; Sharon R. Freeman, Animal Sci. Dept., NC State Univ.; and Matthew H. Poore, Animal Sci. Dept., NC State Univ.
becky_spearman@ncsu.edu

VARIETY SELECTION OF SUMMER ANNUAL GRASSES

D. D. Harmon and D. W. Hancock[1]

Warm-season annual forages have increased in popularity over the last decade, but varieties of forage sorghum (FS), sorghum x sudangrass (SxS), or pearl millet (PM) are often chosen on the basis of convenience and affordability. Although there are many varieties available, it is important to compare their performance in numerous environments and conditions. The goal of this project was to develop a selection tool to help producers determine varieties of FS, SxS, and PM that consistently yield well across multiple environments. Forage yield data from 1998-2014 was collected from the University of Georgia (1998-2014), University of Kentucky (2001-2014), Virginia Tech (2009-2014), and the Noble Foundation (2001-2010). There were 126, 200, and 54 individually identified varieties of FS, SxS, and PM, respectively. The analysis excluded varieties that had been evaluated in fewer than 6 site-year combinations. To standardize across environments, yields were expressed as a percentage of the mean within a given site-year and standardized means were averaged across all available site-years for that variety. Varieties that had standardized mean yields greater than 115% included SS405 (122%), 86S (118%) and Ensile Master (118%) for FS and Fastgrass 5 (120%) and SS 211 (118%) for SxS. No variety of PM was greater than 106%, though both Tifleaf 3 and Elite III achieved this level. Results from this study will help producers identify robust varieties that consistently produce well in diverse environments and conditions.

Deidre Harmon (harmondd@uga.edu) and Dennis Hancock (dhancock@uga.edu), Dept. of Crop and Soil Sciences, University of Georgia, Athens, GA 30605.

EVALUATION OF WARM SEASON ANNUAL FORAGES IN FORAGE-FINISHING BEEF CATTLE PRODUCTION SYSTEMS

D. D. Harmon, D. W. Hancock, R. L. Stewart, Jr., A. M. Stelzleni, C. D. Teutsch and J. R. Segers,

Research is needed to determine if summer annual forage grasses can produce desired cattle gains in a grass-finishing system. The objective of this research was to evaluate and compare four warm-season annual forages for Southeastern forage-finished beef production systems. The experiment was conducted in 2013-2015 at the UGA Beef Research Farm near Eatonton, GA. Pastures (2-acre) were assigned one of four treatments: brown mid-rib sorghum sudangrass (BMR), sorghum sudangrass (SS), pearl millet (PM), or pearl millet plus crabgrass (PMCG; n=4). Thirty-two British-cross beef steers (963±44 kg) were assigned to one of the 16 pastures for forage finishing (n=2). Pastures were rotationally grazed and forage allowance was monitored biweekly and put and take grazing was used to maintain a forage allowance goal of 1500 – 4000 lbs/acre. Forage quality and steer growth measurements were taken on d 0, 34, and 70. Forage mass was greatest from mid-June-late July for the SS and BMR treatments (2450 – 3700 vs. 900 - 2200 lbs/a), but the PM treatments sustained greater yields in late Aug.-Sept. (2300-2700 vs. 1100-2200 lbs/a). Average daily gain was greatest for BMR (2.04 lbs/hd/d) and least for SS and PM (1.65 lbs/hd/d, each), with PMCG as intermediate (1.79 lbs/hd/d). The BMR gave superior animal performance, though producers should account for peak forage production occurring in June-July.

D. D. Harmon (harmondd@uga.edu), D. W. Hancock (dhancock@uga.edu), Dept. of Crop and Soil Sciences, University of Georgia; R. L. Stewart, Jr. (lawtons@uga.edu), A. M. Stelzleni (astelz@uga.edu), J. R. Segers (jacobs@uga.edu), Dept. of Animal and Dairy Sciences, University of Georgia; C. D. Teutsch (cteutsch@vt.edu), Dept. of Crop and Soil Environmental Sciences, Virginia Polytechnic Institute and State University.

COVER CROPS: TO GRAZE OR NOT TO GRAZE-IMPACTS ON SOIL HEALTH

T.R. Higgins

Cover crops are becoming increasingly popular as a means to protect the soil and improve soil health. An appealing aspect of cover cropping for farmers with integrated crop and livestock systems is grazing potential. The question addressed was, does grazing cover crops have an impact on soil health? The three-year study looked at soil health from the perspective of organic matter content change over time and season, soil density, and earthworm populations. Five cover crop treatments (cereal rye alone and cereal rye with purple top turnips, hairy vetch, crimson clover, or rape) and two grazing treatments (graze or no graze) were used. Cover crops were sown in the fall after corn and were grazed in the spring by meat goats. Once grazing finished the fields were sprayed with glyphosate and the corn or soybean crop was planted and any cover crop residue roller crimped. Soil samples were collected in the spring and fall as were soil density and earthworm population readings. These data were compared across cover crop treatments and grazing treatments. Seasonal variation in organic matter content was observed. Earthworm populations tended to be lower in paddocks that were grazed versus those that were not grazed. We expected that the lower earthworm populations observed after grazing in the spring would recover by the fall, but this did not generally happen. It is possible that grazing changes the organic matter profile from one that decomposes more slowly to one the decomposes more rapidly and thus it may be that grazing does not support higher earthworm populations due to the rapid conversion of the plant material into manure.

Todd R. Higgins, Lincoln University, Jefferson City, Missouri, HigginsT2@LincolnU.edu

COMPARISON OF FIVE NONTRADITIONAL LIQUID FERTILIZERS FOR FORAGE YIELD

John Jennings, Kenny Simon, Leo Espinoza, Dirk Philipp, Shane Gadberry, Paul Beck, Paul Ballantyne, and Ronald Lathrop[1]

High commercial fertilizer prices cause many producers to look for lower-cost fertilizer alternatives. The objective of this study was to compare dry matter yield of late summer and fall-grown stockpiled bermudagrass and stockpiled tall fescue treated with five different nontraditional liquid fertilizers. Experiments were conducted at the Southwest Research and Extension center and the Watershed Research and Extension Center in Arkansas. Products compared were Monty's Plant Food, Sea Mineral, Royal Grow, Grasshopper, and AgriGro Foliar Blend. Each liquid product was applied alone at recommended rates and in combination with commercial fertilizer based on soil test recommendations. According to recommendations for stockpiling forage, treatments for stockpiled bermudagrass were applied in August and plots were harvested in October and treatments for stockpiled fescue were applied in early September and plots were harvested in early December. Significant dry matter yield increases were measured for commercial fertilizer for most, but not all species and locations. Liquid fertilizer products compared in this study had no significant effect on forage dry matter yield of stockpiled bermudagrass or stockpiled fescue at four locations. Results indicate that the liquid fertilizer products used under conditions of this study were not effective for improving forage yield or forage quality.

[1]Professor, (jjennings@uaex.edu); Program Associate, (ksimon@uaex.edu); Associate Professor, (lespinoza@uaex.edu); Associate Professor, (dphilipp@uaex.edu); Associate Professor, (sgadberry@uaex,edu); Professor, (pbeck@uaex.edu); Program Associate; Program Associate; University of Arkansas, Division of Agriculture, Little Rock, Arkansas 72204

COMPARISON OF FIVE NONTRADITIONAL LIQUID FERTILIZERS FOR FORAGE NUTRITIVE VALUE

John Jennings, Kenny Simon, Leo Espinoza, Dirk Philipp, Shane Gadberry, Paul Beck, Paul Ballantyne, and Ronald Lathrop[1]

High commercial fertilizer prices cause many producers to look for lower-cost fertilizer alternatives. The objective of this study was to compare nutritive quality (CP & TDN) of late summer and fall-grown stockpiled bermudagrass and stockpiled tall fescue treated with five different nontraditional liquid fertilizers. Experiments were conducted at the Southwest Research and Extension center and the Watershed Research and Extension Center in Arkansas. Products compared were Monty's Plant Food, Sea Mineral, Royal Grow, Grasshopper, and AgriGro Foliar Blend. Each liquid product was applied alone at recommended rates and in combination with commercial fertilizer based on soil test recommendations. According to recommendations for stockpiling forage, treatments for stockpiled bermudagrass were applied in August and plots were harvested in October and treatments for stockpiled fescue were applied in early September and plots were harvested in early December. Forage samples were analyzed with NIR. Significant forage quality increases were measured for commercial fertilizer at most, but not all species and locations. With few exceptions, liquid fertilizer products did not improve forage CP or TDN. Results indicate that the liquid fertilizer products used under conditions of this study were not effective for improving forage quality.

[1]Professor, (jjennings@uaex.edu); Program Associate, (ksimon@uaex.edu); Associate Professor, (lespinoza@uaex.edu); Associate Professor, (dphilipp@uaex.edu); Associate Professor, (sgadberry@uaex,edu); Professor, (pbeck@uaex.edu); Program Associate; Program Associate; University of Arkansas, Division of Agriculture, Little Rock, Arkansas 72204

ESTABLISHMENT OF NATIVE WARM SEASON GRASSES IN ARKANSAS

John Jennings, Kenny Simon, Don Hubbell, Michelle Mobley, Hank Chaney, and Pat Keyser[1]

Native warm-season grasses are tall-growing bunch grasses grown for livestock forage and wildlife cover and include big bluestem, little bluestem, switchgrass, and indiangrass Typical recommendations suggest it takes 2-3 years to get a full stand of native grasses established which makes many producers shy away from planting native grasses. This demonstration tested best available recommendations for establishing native grasses in Arkansas. Three on-farm sites totaling 35 acres were planted in May and June 2014. Two sites were planted in undisturbed killed sod and a third was planted on a tilled, clean seedbed. Existing sod was sprayed in spring multiple times with glyphosate herbicide prior to planting. A seed mix of 10 lbs/acre PLS consisting of 80% big bluestem, 10% little bluestem and 10% indiangrass was planted by no-till drill. Drills were adjusted to plant seed firmly at the soil surface. Fertilizer and lime were applied as needed based on soil test recommendations. Plateau herbicide was applied immediately after planting for summer weed control. Excellent stands were established at all locations with stand counts exceeding 80% by October. Total costs ranged from $337 to $400/acre. Seed cost ranged from 38% to 45% of total costs. Weed control expenses ranged from 32% to 51% of total costs. Weed control proved to be a critical factor in project success based on stand counts and observation of unsprayed areas.

[1]Professor, (jjennings@uaex.edu); Program Associate, (ksimon@uaex.edu); Livestock and Forestry Research Station Director, (dhubbell at uark.edu); Cleburne County extension agent, (mmobley@uaex.edu); Faulkner County extension agent, (hchaney@uaex.edu); University of Arkansas, Division of Agriculture, Little Rock, Arkansas 72204; Director of Center for Grassland Renewal, University of Tennessee, Knoxville, TN

POLYMORPHISMS AT DRD2 GENE AFFECT AVERAGE DAILY GAINS OF COW-CALF PAIRS ON TALL FESCUE PASTURE

B.T. Jones[1], R.L. Kallenbach[1], and C.J. Kojima[2]

Livestock consuming the ergot-like alkaloids produced by tall fescue [*Lolium arundinaceum* (Schreb.) Darbysh.] infected with wild-type *Epichloe coenophiala* display reduced average daily gains. This phenomenon called *tall fescue toxicosis* costs U.S. producers around $1 billion annually. A recently discovered dopamine receptor D2 (DRD2) gene may help cattle tolerate these alkaloids. The objective was to determine the response of cow-calf pairs selected for tolerance or susceptibility to toxicosis when grazing toxic or non-toxic tall fescue pasture. Five cow-calf pairs were rotationally grazed on each of the different tall fescue-endophyte associations which were replicated three times (n=30 cow-calf pairs each year). Body mass of cows and calves was collected along with pasture quality data. Ergovaline concentration of KY-31 tall fescue was greater than for BO+. Our data indicates that tolerant genotyped cows grazing KY-31 raised calves with 54 lbs./head greater 205-d adjusted weaning weights ($P = 0.03$) than susceptible cows on the same forage. Cow genotype had no influence ($P = 0.99$) on 205-d adjusted weaning weight of calves when BO+ was grazed, though the weaning weights were 39 to 93 lbs./head greater than for either genotype grazing KY-31. The DRD2 gene shows promise as a way to mitigate tall fescue toxicosis if non-toxic pastures are not feasible. [1]

---

[1] [1]Graduate Research Assistant, Division of Plant Sciences, University of Missouri, Columbia, MO (bj2mb@mail.missouri.edu); [1]Professor, Division of Plant Sciences, University of Missouri, Columbia, MO (KallenbachR@missouri.edu); [2]Associate Professor, Department of Animal Science, University of Tennessee, Knoxville, TN (ckojima@tennessee.edu).

ORCHARDGRASS DIE-OFF: HOW HARVEST MANAGEMENT AND HEAT STRESS MAY BE CAUSING REDUCED PERSISTENCE IN THE MID-ATLANTIC

G. B. Jones and B. F. Tracy[1]

Orchardgrass is one of the most important hay species in Virginia and the Mid-Atlantic. In recent years, hay producers have noted decreased persistence of their orchardgrass hay stands. Several hypotheses have been proposed to explain the decreased persistence including low cutting heights, soil fertility imbalances, disease, and pest pressure. A literature review and field survey of 53 orchardgrass stands in four states indicated that soil fertility and pest pressure do not appear to be driving the die-off of orchardgrass. The literature on cutting height of orchardgrass shows differing conclusions about how low to cut. We hypothesize that cutting height interacts with weather conditions following harvest to cause reduced persistence. When stands are harvested to a low cutting height, photosynthetic leaf area and the pool of soluble carbohydrates available for regrowth become limited. Also, the amount of shade to the soil surface is reduced creating a hotter environment for the regrowth of the already stressed plants. This effect is compounded in plants already stressed by warmer than optimal temperature and by delaying first cutting into early summer. Slow regrowth during mid-summer further limits carbohydrate synthesis leading to tiller death and decreased stand persistence. Preliminary results of several orchardgrass experiments will be discussed.

---

[1] Graduate Assistant and Associate Professor of Grassland Ecosystem Management, respectively, Department of Crop and Soil Environmental Sciences, Virginia Tech, Blacksburg, VA.

IS FEEDING SEED TO CATTLE THROUGH MINERAL SUPPLEMENTS A VIABLE ALTERNATIVE TO FROST-SEEDING RED CLOVER INTO ENDOPHYTE-INFECTED TALL FESCUE PASTURES?

Paige M. Kennedy[1], Matthew H. Poore,[1] Ben M. Goff[2], and Carrie L. Pickworth[1]

Adding legumes to pastures significantly increases weaning weight, average daily gain, and reproductive performance in cattle grazing endophyte-infected tall fescue. However, various constraints including topography and lack of time, have limited legume establishment efforts. Some producers feed seed to cattle in an attempt to easily sow clovers in their pastures through distribution of seed in fecal matter. A small plot study tested the efficacy of this practice assuming all seeds survived passage. A comparison was made of red clover (RC) establishment in 'KY-31' tall fescue pastures between coated and uncoated seeds, either frost-seeded (FS) or distributed in a fecal pat (F). A dragging treatment (with or without) was applied to quantify impact of this management on improving seedling survival and distribution. Treatments were applied in mid-February and seedlings counted a total of four times from March to May. The uncoated-drug FS treatment yielded the greatest number of seedlings. FS controls had the highest RC seedling densities (79.8, 101.9, 86.4, and 86.1 seedlings/y^2), with little effect of dragging. Dragging significantly improved establishment in F treatments. This study suggests that dragging improves establishment of RC seeds that are present in fecal pats. It is critical to remember that this study removed the impact of bovine digestion, and frost seeding of uncoated or coated seeds may still be the most suitable method for establishment of RC into existing pastures.

[1]College of Agriculture and Life Sciences, North Carolina State University; (pmkenned@ncsu.edu) (mhpoore@ncsu.edu) (clpickwo@ncsu.edu)
[2] Department of Plant and Soil Sciences, University of Kentucky (ben.goff@uky.edu)

PASTURE HERBICIDE RESIDUE EFFECTS ON CRIMSON AND WHITE CLOVER

Robert A. Lane and Mark J. Anderson[1]

To extend the grazing season and provide high quality cool season forage, southern forage producers frequently overseed warm season perennial grass pastures with legume crops in autumn. Producers commonly use herbicides in the summer months to control broadleaf weeds in grass pastures. Those herbicides may cause production issues with fall-planted legumes. In this study, four replicates of each Chaparral™ (2 oz/A and 2.5 oz/A), GrazonNext HL™ (1.5 pt/A and 2.1 pt/A), and Grazon P+D™ (2 pt/A) were applied to a mixed common/Coastal bermudagrass pasture on April 15, May 15, June 15, and July 15, 2013. Dixie crimson clover and Durana white clover were overseeded across all treatment plots on October 15, 2013. Clover density in February, 2014 was significantly reduced in all May, June and July herbicide treatments except Grazon P+D. Clover in nearly all herbicide treatments for all months exhibited at least some degree of phytotoxicity in February compared to the untreated control. The exception was Grazon P+D, which caused less phytotoxic effects than the other herbicide treatments applied in April, May and June. Based on results observed in this study, if a mid-summer herbicide application is made, Grazon P+D would be less likely to contribute to reduced crimson and white clover performance. Under similar conditions, it would be prudent to avoid Chapparal and GrazonNext for summer weed control where fall-planting of legumes is practiced.

[1]Professor and Assistant Professor, respectively, Sam Houston State University
Dept. of Agricultural Sciences and Engineering Technology, Huntsville, Texas. blane@shsu.edu

PRODUCTIVITY OF FORAGE WHEAT AND ANNUAL RYEGRASS FOLLOWING ROUNDUP READY FORAGE SOYBEANS

David Lang, Glover Triplett, David Russell, Beau Varner, Graham Oakley and Ernest Flint[1]

Soybeans provide high quality forage for livestock and wildlife. Utilization of glyphosate tolerant (RR) soybeans provides the opportunity to kill toxic tall fescue and other plants. RR soybeans as a grain type and 'Gamekeeper' were no-till planted in June for two years that previously had tall fescue. Forage wheat and Marshall annual ryegrass were no-tilled planted into the harvested soybean stubble in the late fall of 2013 and 2014. Wheat and ryegrass were fertilized with N at 0, 30 and 60 lbs/acre following each harvest beginning in March. Soy as hay in 2014 was 5100 lbs/acre for the grain type and was 12,000 lbs/acre for Gamekeeper as forage. Soy hay yield in 2015 ranges from 6500 lbs/acre for the grain type and 8000 lbs per acre for Gamekeeper as forage. Protein of the soy hay ranged from 13% for the grain type and 15% for Gamekeeper forage. Digestibility was 67% for grain soy hay and 69% for forage soy hay. Ryegrass and wheat responded to N rates, but their yields were low: 3200 to 5000 lbs/acre for ryegrass and 1500 to 2700 lbs/acre for wheat Wheat harvested as forage was absent in the May harvest and may be a better forage to grow before soybeans. Ryegrass continued to grow into May and would have competed with soybean planting. Delaying soybean harvest until late October or early November resulted in a late planting of ryegrass and wheat. Protein levels in ryegrass and wheat were 18 or 14%, respectively indicating N from soybeans was similar.

[1]Department of Plant and Soil Sciences and Mississippi Cooperative Extension, Mississippi State University dlang@pss.MsState.edu

LONG-TERM ANALYSIS OF FORAGE QUALITY IN MISSISSIPPI

R. Lemus, J.D. Rivera, and K.J. Han[1]

Forage and livestock producers have focused hay production efforts in increasing yield while sacrificing forage quality. This is typically reflected by poor forage intake and lower digestibility. This strategy is causing an increase in cost of supplementation and limiting the need to meeting livestock nutrient requirements. The objective was to analyze forage quality data from samples submitted by Mississippi's forage producers from 2006 to 2014 to the former LSU Forage laboratory. Samples were divided into annual ryegrass (*Lolium multiflorum* L.), bermudagrass (*Cynadon dactylon* L.), bahiagrass (*Paspalum notatum* L.), mixed hay, baleage, and silage. Samples were scored to determine which percent of the sample population met livestock nutrient requirements according to the National Research Council Guidelines. Overall analysis indicated that a great majority of the samples submitted by producers did not meet the nutrient requirements of the livestock classes compared in the analysis. This is an indication that while the animal performance depends on factors such as genetics, animal health, and forage availability; forage quality plays a key role in lowering the cost of supplementation and increasing the return per cow. Forages with lower fiber content (NDF) will allow increase in intake and perhaps better forage utilization. Our goal is to help producers to understand that better forage quality is a function of hay management that might require little additional inputs.

[1]Extension Forage Specialist, Center for Forage Management & Environmental Stewardship, Mississippi State University (Rocky.Lemus@msstate.edu); Animal Scientist, South Mississippi Branch Experiment Station, Mississippi State University (Daniel.Rivera@msstate.edu); Associate Professor, Forage Production/Quality Evaluation, Louisiana State University AgCenter (Khan@agcenter.lsu.edu).

IMPACT OF GRASS/LEGUME MIXTURE IN WINTER GRAZING SYSTEMS

R. Lemus, J.A. White, and J.D. Rivera[1]

The cost of chemical fertilizers has increased exponentially in the last decade, impacting forage production and grazing systems during the winter. There is potential to expand cool-season forage opportunities by evaluating new annual clover varieties under grazing conditions in Mississippi. The purpose of this demonstration was to evaluate how the management of annual ryegrass and legumes in a monoculture or mixed system will benefit conventional vs. sod seeding systems. Two, two-acre paddocks were used for conventional seeding and one two-acre bermudagrass paddock was used for sod-seeding. Each paddock was divided in approximately 0.67-acre strips to establish three treatments. The treatments include annual ryegrass (LoneStar and Marshall) fertilized with commercial fertilizer, annual ryegrass + Frosty berseem clover, and Frosty berseem clover alone. LoneStar provided a slight grazing advantage early in the season with earlier biomass production compared to Marshall, although that advantage was mitigated by the second grazing cycle. The total gain per acre in each system was justified by the amount of forage production and the high forage quality provided by each system. There is a need for replicated studies that will allow us to make statistical comparisons on biomass production, animal performance, and the economic benefits of clovers (cost vs. return) in large scale grazing systems.

[1]Extension Forage Specialist, Center for Forage Management & Environmental Stewardship, Mississippi State University (Rocky.Lemus@msstate.edu); Forage Variety Testing Manager, Mississippi Agricultural & Forestry Experiment Station, Mississippi State University (Joshua.White@msstate.edu); Animal Scientist, South Mississippi Branch Experiment Station, Mississippi State University (Daniel.Rivera@msstate.edu).

GRAZING EVALUATION OF COOL-SEASON GRASSES WITH AND WITHOUT LEGUMES

R.W. McKee, J.M. Johnson, and E.V. Santen[1]

Reducing the length of time that cattle must be sustained on stored forages during the winter by extending the grazing season is a growing interest of cattle producers in the Southeast. This project was conducted to evaluate several of the predominant cool-season forages and forage mixtures for their capability to extend the grazing season. Six cool-season forage treatments were assigned to 0.8-ha paddocks in a completely randomized design with two replications and continuously grazed at a fixed stocking rate of 5 steers/ha (4 steers/paddock). The treatments were as follows: 'Texoma' MaxQ II novel endophyte tall fescue grown in combination with 'Durana' white clover (TF+WC) or treated with 50.5 kg N/ha (TF), 'Nelson' annual ryegrass grown in combination with 'Dixie' crimson clover (RG+CC) or treated with 50.5 kg N/ha (RG), and a mixture of 'Graze King 90' cereal rye and 'Nelson' annual ryegrass grown in combination with 'Dixie' crimson clover (RG+R+CC) or treated with 50.5 kg N/ha (RG+R).

In year one treatments containing tall fescue furnished 75 days of grazing while all other treatments furnished 68 days. In year two treatments containing tall fescue provided 84 days, RG+R and RG+R+CC 57 days, and RG and RG+CC provided 85 days of grazing. In year one steer ADG was greatest on RG+CC and RG+N but did not differ significantly from TF+WC, which was similar to all treatments. In year two steer ADGs were greatest on the annual treatments, which did not differ. Results indicate that crimson clover can mitigate N fertilizer needs for spring grazing on annual ryegrass and annual ryegrass-cereal rye pastures.[1]

---

[1] [1]Graduate Research Assistant; Assistant Professor and Forage Extension Specialist; Professor, Department of Crop, Soil, and Environmental Science, Auburn University, Auburn, AL.

IMPACT OF SHADE ON THE YIELD, NUTRITIVE VALUE, AND BOTANICAL COMPOSITION OF FORAGE MIXTURES FOR SILVOPASTURES

K.M. Mercier, C.D. Teutsch, J.H. Fike, and K.K. Quick[1]

Interest in silvopastures in Virginia and other transition zone states is increasing. Silvopasture is the intentional integration of trees, forages and livestock into an intensively managed system. Forages established in silvopasture systems are subject to different environmental conditions than typical pastures as a result of the changes in microclimate due to presence of trees. These conditions may impact both yield and forage diversity. The objective of this experiment was to evaluate the impact of shade on the botanical composition and yield of a simple, intermediate, and complex cool-season forage mixtures. The experimental design was a random complete block with four replications. Slatted shade structures were used to create three different light levels (25, 50, and 75% full sun) for comparison with a no shade control. In early April, 2015, three different cool-season forage mixtures of varying complexity were planted. Immediately after planting, shade structures were placed over the corresponding plots. Yield at the first and second harvest decreased by more than 35% as shade level increased ($P < 0.03$). Mixture complexity had no effect on yield ($P > 0.59$). The amount of tall fescue, orchardgrass, and red clover in the complex mixture tended to increase as shade increased. All three mixtures had fewer weeds with shade when compared with the no shade control ($P < 0.04$). Shade appeared to hinder the growth of summer annual grasses such as crabgrass, which could enhance initial establishment in spring seedings.

[1]K.M. Mercier, Virginia Tech, Southern Piedmont AREC, Blackstone, VA (kmerc094@vt.edu); C.D. Teutsch, Virginia Tech, Southern Piedmont AREC, Blackstone, VA (cteutsch@vt.edu); J.H. Fike, Virginia Tech, Crop and Soil Environmental Sciences, (jfike@vt.edu); and K.K. Quick, Virginia Tech, Southern Piedmont AREC, Blackstone, VA (kkquick@vt.edu).

EASTERN GAMAGRASS ESTABLISHMENT WITHOUT BREAKING THE BANK

J.I. Morrison and B.S. Baldwin[1]

Eastern gamagrass is a hearty, long-lived perennial warm-season native grass. The species has been utilized for forage, wildlife, bioenergy, riparian buffer, and ornamental purposes. The most valuable and applicable of these purposes is as a forage crop, most effectively utilized as a grazing or hay crop. The major impediments to more widespread use of eastern gamagrass are its hard seed coat and deeply dormant embryo, which delay germination, emergence and establishment. Eastern gamagrass seed units are large, woody cupules that encapsulate the embryo and endosperm. These seed units have been shown to exhibit physical, mechanical and physiological dormancy. The standard practice for treating gamagrass seed prior to planting involves six to eight weeks of cold, moist stratification followed by applications of fungicide and, often, hydrogen peroxide. These practices cause current gamagrass seed prices to range between 15 and \$25/lb, which equates to 120 - \$250/acre for seed. A research study was designed in 2011 to eliminate physiological dormancy from eastern gamagrass via recurrent phenotypic selection breeding. Subsequent germination and selection has led to three established breeding populations representing four cycles of selection. Initial 6-day germination of non-stratified base population seedlot was 0%, while 28-day germination of the same seedlot was < 1%. Significant improvement in germination has been accomplished with mean 14- and 28-day germination of Gen_1 seedlot at 2.91% and 4.33%, respectively, in its second production year and Gen_2 seedlot at 4.71% and 8.67%, respectively, in its first production year.

[1]Research Associate II, Department of Plant and Soil Sciences, Mississippi State University, Starkville, MS (jesse.morrison@msstate.edu); Professor, Department of Plant and Soil Sciences, Mississippi State University, Starkville, MS (bbaldwin@msstate.edu)

RESPONSE OF SMOOTH BROMEGRASS FORAGE YIELD AND QUALITY TO NITROGEN MANAGEMENT ALTERNATIVES

J. L. Moyer and D. W. Sweeney[1]

Smooth bromegrass is a cool-season perennial grass used in pastures and meadows of the northern U.S. Nitrogen (N) availability is often a limiting factor in its production, so this study was conducted to address the effect of N rate, timing, and method of application on bromegrass forage yield and quality at two growth stages. For four years, urea-ammonium nitrate (UAN) was applied at two rates, 75 and 150 lb N/acre/yr, by three N placements: broadcast (BC), dribble (DR), and knife (KN), plus a no-N check. Four fall/late winter N timing combinations were compared: 1) 100% in late fall (F), 2) 67% F, 33% in late winter (S), 3) 33% F, 67% S, or 4) 100% S. Forage was sampled at a growth stage that was early in stem elongation to estimate grazing potential, and the rest of each plot was harvested post-anthesis to measure hay production. Early yield was increased an average of 17% when N rate increased from 75 to 150 lb/acre, while hay yield was increased by 22%. Crude protein (CP) was also increased as N rate increased, but in vitro digestibility (IVDMD) was decreased slightly. Knife (KN) placement generally resulted in less production at the early stage, but hay yield was increased by 19% from KN compared with BC. Crude protein of hay was increased in DR and KN placement, but IVDMD was reduced with KN. Applications split between F and S usually resulted in less early production, but more hay yield than single applications, whereas forage quality estimates were more variable.

[1]Professors, Kansas State University Southeast Agricultural Research Center, Parsons, KS 67357 (jmoyer@ksu.edu and dsweeney@ksu.edu).

ECOLOGICAL SITE DESCRIPTIONS AND THEIR ROLE IN CONSERVATION PLANNING

J. Pate[1]

Looking across any landscape, it is not difficult to recognize that some parts are different from other parts in regard to the plant communities present, especially as to the kinds and amounts of vegetation. To understand this variation across the landscape, we classify these different parts into units called Ecological Sites. An Ecological Site is defined as "a distinctive kind of land with specific characteristics that differs from other kinds of land in its ability to produce a distinctive kind and amount of vegetation". Land inventory, analysis, and resulting management decisions require the knowledge of these individual sites and their relationships on the landscape. The Ecological Site Description (ESD) is the document that contains information about the individual ecological Sites. ESD's provide land managers the information needed for evaluating the land as to suitability for various land-uses, capability to respond to different management activities or disturbance processes, and ability to sustain productivity over the long term. ESD information is presented in four major sections:

1) Site Characteristics - physiographic, climate, soil, and water features
2) Plant Communities – plant species, vegetation states, and ecological dynamics
3) Site Interpretations – management alternatives for the site and its related resources
4) Supporting Information – relevant literature, information and data sources

[1]J. Pate, USDA-NRCS, Fort Worth, TX (Johanna.pate@ftw.usda.gov).

UTILIZING UREASE INHIBITORS FOR STOCKPILING FESCUE

C.D. Penrose, M.A. Landefeld, J.S. McCutcheon[1]

One way to lower the amount of stored feed needed is to field-stockpile forages for fall and winter grazing. Studies have demonstrated that one way to improve the quality and yield of fescue is to apply nitrogen fertilizer (N) such as urea when stockpiling is initiated. The biggest risk when applying urea is not receiving adequate rainfall and volatilizing before it reacts with the soil. Volatilization losses may be reduced if a urease inhibitor such as Agrotain® is used with the urea. The objective of this study was to determine if the addition of a urease inhibitor can increase yield and quality of stockpiled fescue. The replicated study was initiated on 8/8/2014 and plots harvested on 12/3/2014 comparing no N, 46 lb N/acre, and 46 lb N/acre with Agrotain®. Dry matter yields averaged 2369 lb/acre for plots with no N, 3147 lb/acre for plots with 46 lb N/acre, and 3210 lb/acre for plots with 46 lb N/acre with Agrotain®. There was a significant difference ($P \leq 0.05$) in crude protein (CP) with urea and urea with Agrotain® compared to the control plots. CP averaged 6.77% for plots with no urea, 8.53% for plots with 46 lb N/acre and 8.32% for plots with 46 lb N/acre with Agrotain®. One possible explanation for the lack of significant differences between N treatments was unplanned rainfall (0.59”) within 72 hours of the initiation of the study and 0.56” the next day. Agrotain® may be an option if there is doubt about rain after applying urea in warm weather.

[1]Extension Educator, Ohio State University, McConnelsville, OH (penrose.1@osu.edu); Extension Educator, Ohio State University, Mt. Gilead, OH (mccutcheon.30@osu.edu); Educator, Ohio State University, Woodsfield, OH (landefeld.6@osu.edu).

SHEEP PERFORMANCE, GRAZING BEHAVIOR, AND BODY TEMPERATURES IN SILVOPASTURE SYSTEMS

Gabriel Pent & John H. Fike[1]

Silvopasture systems integrate trees with forages for grazing livestock. In some comparisons with open pasture systems, animal gains have been similar or better in silvopastures despite lower forage yield. The objective of this study was to determine whether changes in forage nutritive value or changes in lamb behavior during summer might compensate for lower forage productivity in honeylocust and black walnut silvopasture systems. Tree species had variable effects on forage characteristics and animal behavior, but lamb weight gains were similar across all treatments. Forages in the honeylocust silvopasture systems were more productive and nutritious than forages in the open pasture systems. Forages were least productive and nutritious in the black walnut silvopasture systems. However, greater shading in walnut systems kept the lambs cooler throughout the day. Lambs in silvopasture systems spent more time lying down and distributed grazing events throughout the day. Lambs in the open pastures spent more time standing and concentrated their grazing events into evening hours. Interestingly, although parasites kept lamb weight gains low across treatments, lambs in the open pastures had higher fecal egg counts and lower weight gains.

[1] Graduate Research Assistant, Department of Crop and Soil Environmental Sciences, Virginia Tech; Associate Professor, Department of Crop and Soil Environmental Sciences, Virginia Tech

SHEEP PERFORMANCE, GRAZING BEHAVIOR, AND BODY TEMPERATURES IN SILVOPASTURE SYSTEMS

Gabriel Pent & John H. Fike[1]

Abstract

Silvopasture systems integrate trees for timber or fodder with forages for grazing ruminants. Reports of forage and animal productivity from these systems have been variable. In some cases, animal gains are similar or better despite lower forage yield. We explored the relationship between grazing system, body temperature, grazing behavior, and animal performance in this study. Black walnut (*Juglans nigra*) and honeylocust (*Gleditsia triacanthos*) based silvopasture systems were compared with open pastures in a randomized complete block design (r = 3). Pastures were rotationally stocked with six or seven lambs depending on forage availability. All sheep were weighed and fecal samples from a subset of the lambs were taken every four weeks. Body temperatures and grazing behavior measures were recorded within a replicate within a week, and these measures were taken sequentially within three experimental periods. Ewe lambs (n=3) within each experimental unit were equipped with a vaginal temperature sensor and a wideband audio recording device to detect prehension events. Time lapse cameras documented sheep behavior every 60 seconds. Forage availability measurements taken with a rising plate meter indicated no difference between the honeylocust silvopasture (4400 ± 70 lbs/acre) and the open pasture (4300 ± 70 lbs/acre; P = 0.9916), though forage availability in the black walnut silvopastures (3100 ± 70 lbs/acre) was lower than that within the other treatments (P < 0.0001). Animal gains were highest in the honeylocust silvopastures (27.8 ± 2.12 lbs/period), and were significantly different from the gains of the open pastures (20.8 ± 2.20 lbs/period; P = 0.0254) and the gains of the black walnut silvopastures (18.7 ± 2.22 lbs/period; P = 0.0046), though no difference was found between the latter two systems (P = 0.5173). Observations indicate that sheep in silvopastures had extended foraging periods and, in some cases, lower body temperatures.

Introduction: Silvopasture management entails the intentional integration of livestock, forages, and trees in a single system. Previous research has shown that tree species have differential effects on pasture composition; forage yield and nutritive value responses can be positive or negative depending on tree age, stand density, and slope position, among other factors (Buergler et al., 2005; Buergler et al., 2006). While deciduous silvopastures may differ from open pastures in terms of forage characteristics, these responses do not necessarily track differences in animal performance (Peri et al., 2001; Lehmkuhler et al., 2003; Kallenbach et al., 2006).

Recent research (Fannon-Osborne, 2012) with lambs grazing in walnut- and honeylocust-based silvopasture systems suggests animal performance is comparable to

[1] Graduate Research Assistant, Department of Crop and Soil Environmental Sciences, Virginia Tech; Associate Professor, Department of Crop and Soil Environmental Sciences, Virginia Tech

that from lambs grazing open pastures, even when forage yield is reduced. However, the mechanisms behind these responses have not been clearly defined. Some data suggest that increased forage nutritive value compensates for lower forage mass in silvopastures (e.g. Kallenbach et al., 2006), but lower soluble carbohydrates (Buergler et al., 2006) and only moderate, variable responses in terms of fiber digestibility (Fannon-Osborne, 2012) challenge this idea. Altered animal behaviors – such as grazing time, standing, and lying – and consequences to energy expenditure may thus be more important drivers of the similar animal gains observed between open and silvopasture systems.

Animal behavior in silvopasture has received limited study. Heat load may change activities and intensify stresses experienced by animals in open pastures, thus increasing time and energy spent in behaviors to stabilize body temperature. Ambient temperatures are lower and less variable in silvopasture systems; thus, animals may experience more time with conditions suitable for grazing and increase dry matter intake (DMI) (Mitloehner & Laube, 2003). Distinguishing between reduced energy needed for maintenance vs. greater opportunity for grazing in pasture systems is challenged by the limited tools for monitoring grazing animals, and current methods involve time-intensive observations.

The purpose of this project was to integrate information on forage quantity and quality with spatial and temporal information on grazing behavior and body temperatures to understand the effects of silvopasture dynamics on animal performance.

<u>Materials and Methods</u>: The 12-week grazing study was performed at the Whitethorne Agroforestry Demonstration Center at Virginia Tech's Kentland Farm (Blacksburg, VA). Trees for the silvopasture treatments were planted in 1995 and thinned to a final density in 2012 (about 25 stems per acre). Black walnut (*Juglans nigra*) and honeylocust (*Gleditsia triacanthos*) trees were used to create two silvopasture systems treatments for comparison with open pastures. Each system is 0.67 acres, replicated three times in a randomized complete block design. Pastures and silvopastures were divided into eight subpaddocks for rotational stocking management. The honeylocust silvopasture and open pasture systems were stocked with 7 lambs while the black walnut silvopasture was stocked with 6 lambs. Lambs were weighed with a portable scale for two consecutive days at initiation and conclusion of the study and once each period. Fecal samples from three lambs within each experimental unit were taken at each weigh date to assess for *Haemonchus contortus* infection rates. FAMACHA scores from every lamb were taken every two weeks after the first period and any lamb scoring a 4 or higher after the first period or a 3 and higher after the second period was dewormed with levamisole.

Forage mass was estimated using a rising plate meter before and after every rotation interval; a regression curve was made by double-sampling at every other measurement interval. Samples for nutritive value estimates were collected by hand plucking forage at every other measurement interval. Nutritive value measures were determined by near infrared spectrometry. Vegetation composition was determined every period using a six-class Daubenmeyer approach.

Three sheep in each experimental unit within a block were fitted with a Roland R-05 recorder (Roland Corporation, Los Angeles), which recorded WAV files at a 16 bit resolution and 48 kHz sampling rate, and a Sennheiser ME-2 lavalier microphone (Sennheiser Electronic Corporation, Germany). The devices remained on nine animals for one day before being transferred to sheep in another block the following week for three consecutive weeks of measures in each period. After post processing to remove low frequency noise, prehension events were detected by SIGNAL and GRASS software (Engineering Design, Berkeley). Software parameters were calibrated by determining bite events from the synchronized video and audio files.

Ewe vaginal temperatures were measured by implanting a DST micro-T temperature logger (Star-Oddi, Iceland) into a blank controlled internal drug release device (Zoetis, Florham Park, NJ). The three sheep fitted with the acoustic devices were also equipped with the vaginal temperature loggers for three days in a week.

Moultrie D-500 trail cameras (EBSCO Industries, Inc., Birmingham, AL) were set to visually encompass the entire subpaddock containing the sheep with recorders, capturing images every 60 seconds. Before sampling, the sheep of interest were marked with pink, orange, or blue paint. The photos were processed sequentially by recording the behavior (standing up, lying down, grazing) and shade utilization (in the shade, in direct sunlight) of each sheep by minute.

PROC MIXED in SAS (SAS Institute, Inc., Cary), using a repeated measures model with unstructured variance, was used to conduct an ANOVA of treatment effect on average daily gains (ADG) and fecal egg counts (FEC) of the lambs and total animal gain from each experimental unit. PROC MIXED was used to conduct an ANOVA of treatment effect on forage mass predictions from the rising plate meter and total digestible nutrients (TDN) of orchardgrass (*Dactylis glomerata*) and mixed grass nutritive value samples for each experimental unit.

<u>Results and Discussion</u>: Lambs in the honeylocust silvopastures gained more than lambs in the open pastures ($P = 0.0206$); ADG of lambs in the black walnut silvopastures did not differ from the ADG of lambs in the open pastures or silvopastures ($P \geq 0.25$) (Table 1). Significant treatment by period interaction effects on ADG ($P=0.0191$) likely reflects increasing *H. contortus* loads over time.

Due to the lower stocking rate of the black walnut silvopastures, it is more informative to evaluate animal gains by system rather than by individual animal performance. The gain per experimental unit of the honeylocust silvopastures was significantly higher than that of the open pastures ($P = 0.0254$) and the black walnut silvopastures ($P = 0.0046$). Despite lower stocking rates, gain per experimental unit of the black walnut silvopastures was not significantly different from the gain per experimental unit of the open pastures ($P = 0.5173$).

FAMACHA scores were taken every other week beginning after the first period to determine which lambs should be treated with levamisole. However, lamb anemia due to

H. contortus infection depressed weight gains throughout the study. No treatment effects on FEC were observed, although FEC were numerically lowest in black walnut silvopastures (Table 1).

Throughout the season, forage availability was consistently lowest in the black walnut silvopastures ($P < 0.0001$; Table 1). As a result, the stocking rate of the black walnut silvopastures was kept one animal lower (~55-65 lbs) per experimental unit. No difference was found in forage availability between the honeylocust silvopastures and the open pastures ($P = 0.8709$).

TDN levels in orchardgrass did not differ among the open pasture and silvopasture treatments ($P = 0.6910$; Table 1). However, the TDN of the mixed sward was lower ($P = 0.0055$) in black walnut silvopastures compared with honeylocust silvopastures due to the increased presence of nimblewill (*Muhlenbergia schreberi*) in the walnut system. TDN levels of mixed swards in open pastures did not differ ($P = 0.2919$) from those within the black walnut silvopastures but tended ($P = 0.0721$) to be lower from those within the honeylocust silvopastures.

Ewes in the open pastures had the largest diurnal change in temperatures, with large temperature gains between early morning and late evening and rapid decline late in the day. Body temperatures of the open pasture ewes tracked more closely with the temperature humidity index (THI), presumably because of greater exposure to ambient conditions without tree modulation effects. Unexpectedly, the ewes in the honeylocust silvopasture treatment had the highest peak body temperatures in July and August.

The data to document ewe behavior for this paper were gathered from time lapse photos recorded in August. Ewes in open pastures spent more time standing up than lying down (Table 2). This behavior may reflect efforts to dissipate heat by exposing higher proportions of body surface area to wind. Ewes in the open pasture were able to access to shade around 16:30 as shade from bordering trees covered the experimental unit.

Out of the 28 prehension events counted manually from the video files in a given segment of time, 26 were also detected automatically by GRASS. Using the parameters from this calibration, the acoustic grazing data were analyzed for one date (August 4, 2015) and one ewe per treatment within that date. Grazing time as determined by manual classification of the time-lapse images correlated well with prehension occurrences. The time-lapse images and the acoustic results indicate that the ewe in the open pasture system spent more time grazing with a higher bite rate than ewes in the other systems.

These data indicate that although tree species have variable effects on forage productivity and nutritive value in deciduous silvopastures, animal performance can be equal or better than that in traditional open pasture systems. The environment in silvopastures moderates animal behavior and body temperature, potentially improving animal health and well-being. In turn, adding trees to pastures offers opportunity to increase the overall productivity of the land base.

Literature Cited:

Buergler, A. L., Fike, J. H., Burger, J. A., Feldhake, C. R., McKenna, J. A., & Teutsch, C. D. 2005. Botanical composition and forage production in an emulated silvopasture. Agronomy journal 97(4): 1141-1147.

Buergler, A. L., Fike, J. H., Burger, J. A., Feldhake, C. M., McKenna, J. R., & Teutsch, C. D. 2006. Forage nutritive value in an emulated silvopasture. Agronomy journal 98(5), 1265-1273.

Fannon-Osborne, A. G. 2012. Hair sheep production in temperate, deciduous Appalachian silvopastures (Master's thesis).

Kallenbach, R. L., Kerley, M. S., & Bishop-Hurley, G. J. 2006. Cumulative forage production, forage quality and livestock performance from an annual ryegrass and cereal rye mixture in a pine walnut silvopasture. Agroforestry Sys. 66(1): 43-53.

Lehmkuhler, J. W., Felton, E. E. D., Schmidt, D. A., Bader, K. J., Garrett, H. E., & Kerley, M. S. 2003. Tree protection methods during the silvopastoral-system establishment in midwestern USA: Cattle performance and tree damage. Agroforestry Sys. 59(1): 35-42.

Mitloehner, F. M., & Laube, R. B. 2003. Chronobiological indicators of heat stress in *Bos indicus* cattle in the tropics. J. Anim. Vet. Advances 2:654-659.

Peri, P. L., Varella, A. C., Lucas, R. J., & Moot, D. J. 2001. Cocksfoot and lucerne productivity in a *Pinus radiata* silvopastoral system: a grazed comparison. New Zealand Grassland Association.

Table 1: Performance parameters by treatment.

Treatment	Black walnut silvopasture	Honeylocust silvopasture	Open pasture
Average daily gain (lbs/day)	0.12 ± 0.012	0.14 ± 0.011	0.10 ± 0.011
Gain per experimental unit (lbs/period)	18.7 ± 2.22	27.8 ± 2.12	20.8 ± 2.20
Fecal egg count (eggs/g)	3700 ± 738	4140 ± 728	4600 ± 728
Forage availability (lbs/acre)	3100 ± 70	4400 ± 70	4300 ± 70
Total digestible nutrients in orchardgrass sward (%)	66 ± 0.72	67 ± 0.72	66 ± 0.72
Total digestible nutrients in mixed grass sward (%)	67 ± 0.70	70 ± 0.70	68% ± 0.70

Table 2: Behavioral characteristics of ewes by treatment.

Activity	Black walnut silvopasture	Honeylocust silvopasture	Open pasture
Standing (% of total)	7	5	33
Lying down (% of total)	34	44	6
Grazing (% of total)	59	51	61
Shade utilization (% of total)	99	95	34

STOCKING RATE AFFECTS NET INCOME PER ACRE AND RISK MANAGEMENT: A WEST VIRGINIA CASE STUDY.

E.B. Rayburn[1]

A cow-calf operation was destocked during a major drought. Over the following four years, the farm was gradually restocked from 22 to 32 cows/90 acres. This acreage provided all the grazing and winter feed for the herd. As stocking rate increased, calf sale weight decreased from 655 to 569 lbs./head (average for steers and heifers; 90-97% calf survival). The linear decrease in production per head as stocking rate increased was similar to that documented by numerous research studies. Production/acre was calculated as the product of weight/head and number of animals sold, divided by acreage. When the relation of gain/head and gain/acre were extrapolated past 40 cows, maximum calf production/acre was predicted to be achieved with 45 cows. Prior to destocking, the farm had stocked 40 cows. When cash costs per cow and calf prices were used to calculate net cash income vs. stocking rate, the economic optimal stocking rate was reduced to 35 cows. However, if the stocking rate was reduced an additional 15% to 30 cows, net income would only decrease 2% and significantly reduce risk from drought. Annual forage yield in the region varies with 2-years out of 3 having yields within ±25-30% of average. For this farm, stocking cows at 85% of the economic optimum would reduce net income by 2%, but forage production would be adequate for the herd 70% of the time. This case study demonstrates the value in using stocking rates below the economic optimum, which has low impact on net income, and as a management tool reduces the risk of drought having a negative impact on production.

[1]Extension Specialist, West Virginia University, Morgantown, WV 26506 (erayburn@wvu.edu).

FACTORS AFFECTING PERSISTENCE OF SOUTHEASTERN WILDYE UNDER RIGOROUS HARVEST REGIMES

J.D. Richwine[1], J.I. Morrison[2], and B.S. Baldwin[3]

Harvest frequency and timing impact plant persistence. Abiotic stress from continual harvesting or harvesting above-ground biomass at inappropriate heights often plays a role in increased plant mortality. A study was established in Starkville, MS, in 2013 using a North American native cool-season grass, southeastern wildrye, and harvested for two years during the spring of 2014 and 2015. Plots were fertilized with 0, 120, 180, and 240 lbs N/acre and harvested to a 4-in stubble height using a Ferris® zero turn mower either once (every 112 days), twice (every 56 days), three times (every 37 days), or four times (every 28 days) during the 112 day season. After the final harvest of all plots in 2015, plants counts were taken within a random 3ft^2 within each plot. There was a significant difference among harvests per season with respect to mean number of plants per plot. One harvest during the growing season had significantly higher mean plants per plot than any other treatment. Further research needs to be conducted to assess the impact of various cutting heights and number of harvests per season to determine the effects they would have on southeastern wildrye as compared to a 4-in cutting height.

[1]Graduate Research Assistant, Dept. Plant and Soil Science, Mississippi State University, Starkville, MS (jdr211@msstate.edu); [2]Research Associate II, Dept. Plant and Soil Science, Mississippi State University, Starkville, MS (jim46@msstate.edu); [3]Professor, Dept. Plant and Soil Science, Mississippi State University, Starkville, MS (bsb2@msstate.edu).

SYSTEMS FOR THE CONVERSION OF BERMUDAGRASS TO SWITCHGRASS OR NATIVE GRASSES

J.K. Rogers, J.T. Biermacher, and A. Biedenbach[1]

Bermudagrass is an aggressively spreading warm season perennial grass that is a popular forage in the Southern United States due to its high yield potential and persistence. Disadvantages of bermudagrass are that its production and nutritive value are influenced by nitrogen fertilizer rate and because it is often managed as a monoculture, it provides little wildlife value. Switchgrass, a native warm season perennial, produces grazeable forage mass three to four weeks earlier than bermudagrass. Stocker cattle gain is considered good on spring switchgrass production. Native grass mixtures (little and big bluestem, indiangrass, switchgrass, and green sprangletop) also improve landscape diversity and wildlife habitat. Native grass mixtures are grazed by cattle but not to the intensity of bermudagrass without sacrificing persistence. Managed correctly, switchgrass and native grass require lower inputs of herbicides and fertilizer compared to bermudagrass. Converting established bermudagrass to switchgrass or nativegrass is challenging and difficult. This study evaluated six no-till and six-tillage systems with factorial combinations of preparation time prior to planting and cover crops to convert bermudagrass to switchgrass or nativegrass. The study was repeated twice at two locations and yields were collected after frost for two years. Nativegrass and switchgrass yields were improved with tillage systems compared to no-till systems. Within the tillage treatments, longer preparation times (11 or 19 months) tended to improve grass yields regardless of the use of a cover crop.

[1]Assistant Professor, Pasture and Range, The Samuel Roberts Noble Foundation, Ardmore, OK (jkrogers@noble.org); [2]Center for Economic Information & Analysis Manager, (jtbiermacher@noble.org); [3]Intern, Center for Economic Information & Analysis.

INOCULATION – IS THERE A LINK TO LEGUME PRODUCTIVITY?

J. Rupert[1]

While the use of legumes in forage and cover crop applications is generally accepted as being beneficial in increasing forage quality, forage tonnage, and soil health, the actual correct method of establishing legumes is not consistently practiced. Specifically, the correct application of rhizobium, commonly referred to as inoculation, is a practice which most farmers utilize. However, this practice is not always conducted correctly and is sometimes not even applied, especially on some species where the use of inoculant has not always been deemed necessary. The question we need to consider is this: Is there a link to correct inoculation and legume productivity? We conducted two studies of winter field peas, a species that often times is sown without inoculant. The first study provides a visual example of the difference between both forage growth and root growth of raw and inoculated/coated winter pea. In this 5-gallon bucket study, two levels of inoculant were compared to raw seed. The differences are visibly clear – the inoculated entries had significantly more top growth and both fuller and longer roots as compared to the raw entry. The second study presents forage tonnage data from Luling, TX where two cutting were taken of various legumes including the same variety of winter pea enter as both a raw and inoculated/coated. The most telling data is expressed in the second cutting tonnage where the inoculated/coated entry produced over twice as much dry matter as the raw entry. This raises the question: Are we missing out on the maximum potential of our legumes by either improperly or even skipping the essential step of correctly inoculating our legumes?

[1]Marketing Dev. Manager, Smith Seed Services, Greenville, SC (jruper@smithseed.com).

NATIVE WARM-SEASON GRASSES: CURRENT RESEARCH AND EXTENSION TRAINING IN EAST-CENTRAL MISSISSIPPI

[1]J. B. Rushing and [2]M. Thornton

Several projects related to the establishment and management of native warm-season grasses (NWSG) have been undertaken at the Coastal Plain Branch Experiment Station (CPBES) in Newton, MS. This station focuses on the sustainable production of forages, row crops, and timber while simultaneously creating wildlife habitat and improving soil and water quality. Part of this focus concentrates on the use of NWSG as a drought tolerant, low-input forage source and a component of grassland habitat for a suite of insects, grassland birds, and small mammals. In order to promote the use of these species, several projects have been initiated to validate their implementation into agro-ecosystems. Examples of these projects include: a grazing experiment comparing single and mixed NWSG stands to bahiagrass, a nitrogen and harvest management study comparing mono- and polycultures of NWSG to traditional perennial grasses (bermudagrass, bahiagrass, and tall fescue), a herbicide tolerance study on the native cool-season grass southeastern wildrye, a nitrogen and annual clover study with southeastern wildrye, and a stockpiling study comparing tall fescue and southeastern wildrye beginning in the fall of 2016. Along with these research projects, several extension related activities have been hosted at CPBES to educate agency personnel and the general public on NWSG. These include in-service training events, producer field days, and the Mississippi Forage and Grassland Council Annual Conference held in November 2015. These projects and events have generated a strong interest in the use of NWSG as a forage and for wildlife habitat. Data collection and analysis from research conducted at CPBES will hopefully promote NWSG adoption onto private and public lands of east-central Mississippi.

[1]Assistant Research/Extension Professor; Coastal Plain Branch Experiment Station; Mississippi State University
[2]Graduate Student; Department of Plant and Soil Sciences; Mississippi State University

FORAGES AND GRASSLANDS IN THE PACIFIC NORTHWEST

G.E. Shewmaker[1], D.B. Hannaway[2], and S. Fransen[3]

Forages support livestock industries and rank second among all farm commodities for on-farm cash value in the Pacific Northwest (PNW) states of Oregon, Washington, and Idaho. Eight percent of cropland is classified as pasture for a total of 660,330 acres. Ranking of states for harvested forage (hay and silage) production in the U.S. was 8th for Idaho, 18th for Washington, and 24th for Oregon (USDA-NASS Survey data 2014). There are 18.5 million acres of permanent pasture and rangeland in addition to cropland and woodland pasture area under private ownership, and 35.6 million acres of rangeland under public management. The proportion of total forage land area classified as rangeland is 9% in Washington, 15% in Oregon, and 41% in Idaho. Sixty-two percent of Idaho's land area produces forage, including approximately 47% of the state's acreage suitable for crop production. In addition, domestic and wild animals graze some forest land. Forages are an essential component of livestock production, making up nearly 80% of the diet of the 4.5 million cattle in the PNW. Forage crops in the PNW such as alfalfa (Lucerne; *Medicago sativa* L.) and other hay crops were valued at $2.6 billion, 12% of total U.S. forage value for 2014. Forage, pasture, and range lands in the PNW are vital to sustainable agriculture in the region.

[1]Professor and Extension Forage Specialist, University of Idaho, Kimberly Research and Extension Center, 3806 N 3600 E, Kimberly, Idaho; email: gshew@uidaho.edu; [2]Professor, Forage Research, Extension, and Teaching, Oregon State University, Corvallis 97331, email: david.hannaway@oregonstate.edu; [3]Associate Professor, Forage and Extension Agronomist, Washington State University, Irrigated Agriculture Research and Extension Center, Prosser, WA.

FEEDING STRATEGY AND PASTURE QUALITY RELATIVE TO NUTRIENT NEEDS OF GRAZING DAIRY COWS IN THE NORTHEASTERN U.S.

K.J. Soder[1], A.N. Hafla[1], A.F. Brito[2], R. Kersbergen[3], F. Benson[4], H. Darby[5], and M. D. Rubano[1]

The number of grazing dairies has been on the rise in the last 20 years, particularly in the northeastern U.S. The objectives of this study were to determine the proportions of pasture samples in the northeastern U.S. that meet minimum energy, protein, and mineral requirements of lactating dairy cows and how supplemental feeding impacts nutrient requirements and milk production. Pasture samples were collected from 14 dairy farms in NY, PA, VT, ME and NH over 3 grazing seasons (2012 to 2014) and analyzed for nutritional composition relative to nutrient needs for lactating dairy cows. 86% of pastures did not meet energy requirements for lactating dairy cows, while 21% of samples did not meet protein requirements. Calcium, phosphorus and sulfur were inadequate in roughly 33% of samples, while magnesium and potassium were in excess of dietary requirements. Milk production was higher (45 lb/day) for farms feeding greater levels of supplemental feed, compared with farms feeding no grain/all pasture (37 lb/day). While supplemental feed will increase milk production, it is important to recognize that markets currently exist that pay premiums for value-added milk products (e.g., grass-fed milk) which may negate the economic benefits to supplementation.

[1]Research Animal Scientists, USDA-ARS, University Park, PA (Kathy.Soder@ars.usda.gov; Aimee.Hafla@gmail.com; Melissa.Rubano@ars.usda.gov); [2]Associate Professor, Dept. of Biological Sciences, University of New Hampshire (Andre.Brito@unh.edu); [3]Extension Professor, Dept. of Animal and Veterinary Sciences, University of Maine (Richard.Kersbergen@maine.edu); [4]Small Dairy Extension Educator, Cornell University (afb3@cornell.edu); [5]Associate Professor of Agronomy, University of Vermont (Heather.Darby@uvm.edu).

APPLICATION OF A RISING PLATE METER TO ESTIMATE FORAGE YIELD ON DAIRY FARMS IN PENNSYLVANIA

K.J. Soder, A.N. Hafla, and M. D. Rubano[1]

Accurately assessing pasture forage yield is necessary for producers who want to budget feed expenses and make informed pasture management decisions. The rising plate meter (RPM) is quicker and easier than clipping and weighing samples; however, the RPM requires frequent calibrations and default equations provided by manufactures may be unreliable. Therefore, the objective of this study was to evaluate the ability of a RPM to accurately estimate pasture forage yield on dairy farms. Three Pennsylvania dairy farms utilizing rotational grazing were used in this study. Forage mass was estimated in each pasture one day prior to grazing between August 23 and November 16, 2012 by taking 15 clippings (3' x 4") in the same area the RPM was used (n = 60-75 clips per farm). Visual estimates of botanical composition were made within each pasture to assess species diversity. Equations for estimating pasture forage mass were determined by regressing measured dry matter yield on the corresponding RPM value to produce a linear equation. Manufacturer equations had the greatest standard error (38% of measured forage yield) and lowest R^2 (0.49). The equation considering season had the lowest standard error (9%) and highest R^2 (0.76). Our results indicate that manufacturer equations are unreliable for estimating forage yield on farms in Pennsylvania in pastures with multiple forage species. Calibration equations should be developed and adjusted for each season on each farm to insure the greatest accuracy.

[1]Research Animal Scientists, USDA-ARS, University Park, PA (Kathy.Soder@ars.usda.gov; Aimee.Hafla@gmail.com; Melissa.Rubano@ars.usda.gov);

THE EFFECTS OF UTILIZE AND/OR HYDRA-HUME ON YIELD AND PLANT COMPOSITION OF BERMUDAGRASS HAY

W.A. Storer[1], B.L. Ivey, and B. Chung

Utilize™ and Hydra-Hume™ (Helena Chemical Co.) are two products to be added to a standard fertilization program for improved production. A study was performed to evaluate the effects of application of these products alone or in combination on yield and plant composition of bermudagrass from two separate production programs. In April of 2015, treatments (Hydra-Hume 20 lbs/acre + Utilize 1 pt/acre, Utilize 1 pt/acre, Hydrahume 20 lb/acre, and control) were applied at two locations (Kinder, LA= improved Alicia/Tifton 85 Bermudagrass on sandy silt and Welsh, LA= low-input Jiggs Bermudagrass on clay loam). Yield and plant composition were measured 30 and 60 d after application. At 60 d, plots were scanned with Green Seeker® hand held sensor. At day 30, treatments with Utilize™ increased crude protein percentage at both locations. Treatments with Hydra-Hume™ improved crude protein percentages at the Kinder location with the combined treatment producing a synergistic effect on yield and crude protein. On day 60, no significant differences could be attributed to treatments. In conclusion, Utilize™ produced a transient increase in plant protein that was similar for intensively-managed bermudagrass on sandy soil and low-input Bermudagrass on heavy soil. Hydra-hume™ was more effective at the Kinder location where soils were sandy. Utilize™ and Hydra-Hume™ improved hay quality, but production parameters and soil dynamics play an important role in their efficacy.

[1]Associate Professor, McNeese State University, Harold and Pearl Dripps Department of Agricultural Sciences, Lake Charles, LA (wstorer@mcneese.edu); Graduate Student, McNeese State University, Lake Charles, LA (blivey_sfasu@yahoo.com); Associate Professor, McNeese State University, Lake Charles, LA (bchung@mcneese.edu).

EFFECT OF ENSILING ON ANTI-PARASITIC PROPERTIES OF SERICEA LESPEDEZA

T.H. Terrill , E. Griffin[1], D.S. Kommuru[1], J.E. Miller[2], J.A. Mosjidis[3],
M.T. Kearney[2], and J.M. Burke[4]

Due to its excellent anti-parasitic properties, sericea lespedeza (SL; Lespedeza cuneata), a low-input, warm-season perennial legume well-adapted to the southern and eastern United States, has potential as a natural alternative to synthetic anthelmintics in controlling gastrointestinal worms and coccidia in goats and sheep. Sun-drying SL for hay does not reduce its anti-parasitic properties, but making good quality hay can be challenging because of rapid drying and leaf loss during baling. Ensiling fresh-cut SL would preserve the leaves, but effects of the fermentation process on the anti-parasitic efficacy of this forage are unknown. Therefore, a study was designed to determine the effects of feeding ensiled SL on gastrointestinal nematode (GIN) and coccidia infections in young goats. Both SL hay and SL silage reduced ($P < 0.05$) GIN eggs and coccidial oocysts in feces of the kids compared with others given a control (bermudagrass; Cynodon dactylon) hay diet. Ensiling SL did not reduce its anti-parasitic properties compared with SL hay and may be an effective way to capture more leaves while maintaining the anti-parasitic efficacy of this crop. This will give farmers greater flexibility in the use of this forage as a natural dewormer.

[1]Associate Professor in Animal Science, Graduate Student in Animal Science, Research Assistant, respectively, Fort Valley State University, Fort Valley, GA 31030; [2]Interim Associate Dean for Research and advanced Studies, Biostatistician and Instructor, respectively, Department of Pathobiological Sciences, School of Veterinary Medicine, Louisiana State University, Baton Rouge, LA 70803; [3]Professor, department of Agronomy and Soils, Auburn University, Auburn, AL 36849; [4]Research Animal Scientist, USDA, Agricultural Research Service, Dale Bumpers Small Farms Research Center, 6883 South State Highway 23, Booneville, AR 72927.

EFFECT OF ENSILING ON ANTI-PARASITIC PROPERTIES OF SERICEA LESPEDEZA

T.H. Terrill[1], E. Griffin[1], D.S. Kommuru[1], J.E. Miller[2], J.A. Mosjidis[3], M.T. Kearney[2], and J.M. Burke[4]

Abstract

A study was designed to determine the effects of feeding ensiled sericea lespedeza (SL; *Lespedeza cuneata*) on gastrointestinal nematode (GIN) and coccidia (*Eimeria* spp.) infections in young goats. A 28-day confinement study was completed with 36 nine-month-old Spanish bucks (53.8 ± 8.1 lbs) fed 3 diets consisting of SL silage, chopped SL hay, or ground bermudagrass (BG; *Cynodon dactylon*) hay (70% of the diet) and a grain mixture formulated to balance dietary protein and energy (30% of the diet). Fecal and blood samples were collected every 7 days to determine GIN eggs per gram (EPG), coccidial oocysts per gram (OPG; fecal samples), and packed cell volume (PCV; blood). Both SL diets significantly reduced ($P < 0.05$) EPG and OPG compared to control (BG diet) animals, with greater effects on the OPG numbers. The SL hay diet was slightly more effective against GIN and coccidia than the SL silage, but the differences were not significant. There was no treatment effect on PCV values. As leaf loss during hay making can be substantial with SL, cutting and ensiling at a higher moisture level may be an effective way to capture more leaves while maintaining the anti-parasitic efficacy of this crop.

Introduction: Infection with gastrointestinal nematodes (GIN) and coccidia (*Eimeria* spp.) remains one of the primary constraints to sustainable production of small ruminants world-wide, and with resistance to synthetic anthelmintics now reaching epidemic proportions in sheep and goat GIN, alternative (non-synthetic) control strategies are being explored. One of these is using sericea lespedeza (SL; *Lespedeza cuneata*) as a nutraceutical plant in the diet of ruminants. This warm-season perennial legume contains a high concentration of 'prodelphinidin-type' tannins, which have been associated with excellent efficacy against *Haemonchus contortus*, a highly pathogenic blood-feeding GIN. With growing interest in SL as a natural anthelmintic, particularly by organic livestock producers, a number of studies have been completed documenting the anti-parasitic efficacy of SL in fresh (grazed) and dried (hay, leaf meal, pellets) forms in both sheep (Lange et al., 2006; Burke et al., 2012a, b) and goats (Min et al., 2004; Shaik et al., 2006; Terrill et al., 2007; 2009). While the anti-parasitic value of ensiling condensed tannin (CT)-containing forages has been reported in a few species (Heckendorn et al., 2006), this has not been evaluated for ensiled SL. The objective of the current investigation was to determine the *in vivo* efficacy of ensiled SL against GIN and coccidia in young goats.

---

[1]Associate Professor in Animal Science, Graduate Student in Animal Science, Research Assistant, respectively, Fort Valley State University, Fort Valley, GA 31030; [2]Interim Associate Dean for Research and advanced Studies, Biostatistician and Instructor, respectively, Department of Pathobiological Sciences, School of Veterinary Medicine, Louisiana State University, Baton Rouge, LA 70803; [3]Professor, department of Agronomy and Soils, Auburn University, Auburn, AL 36849; [4]Research Animal Scientist, USDA, Agricultural Research Service, Dale Bumpers Small Farms Research Center, 6883 South State Highway 23, Booneville, AR 72927.

Materials and Methods: 'AU Grazer' SL was harvested from a well-established (9-year) pasture at the Fort Valley State University Agricultural Research Station in Fort Valley, GA, in August, 2014. The forage was cut and chopped to approximately 0.5 inch lengths using a forage harvester. Half the chopped material was immediately packed into 35 gallon black plastic trash bags and tied after manually removing as much air as possible. The bags were then placed inside a second bag, tied, and stored in a barn to allow the bagged material to ensile. The other half of the chopped material was spread out on 100 x 25 foot plastic sheeting and sun-dried for 72 hours, with daily turning to facilitate uniform drying. The dried material was then collected and placed in 55 gallon clear plastic trash bags for storage.

In April and May, 2015, a 28-day confinement feeding trial was completed using 36 nine-month-old intact male Spanish kids (53.8 ± 8.1 lbs) fed 3 isonitrogenous and isocaloric diets consisting of 70% roughage and 30% grain mixture. The roughage portion of the treatment diets consisted of SL silage, SL hay, and ground bermudagrass (BG; *Cynodon dactylon*) hay. The animals were assigned to treatment groups based upon initial fecal egg counts (FEC) and fed their ration in round feed troughs with bars designed to prevent the animals from standing in the troughs. The kids were fed in pens (3 goats per pen) with their rations provided at 3.5% of body weight. All animals were provided *ad libitum* clean water throughout the trial using automatic nipple waterers.

Fecal and blood samples were collected from individual animals on Days 0, 7, 21, and 28 of the trial. Fecal samples were analyzed to determine GIN eggs per gram (EPG) and coccidial oocysts per gram (OPG) using a McMaster technique (Whitlock, 1948). Blood samples were collected into 3 mL EDTA vacutainer tubes and analyzed for packed cell volume (PCV) using a microhaematocrit centrifuge and reader. FAMACHA© scores (van Wyk and Bath, 2002) were also taken on individual animals every 7 days. Animal weights were taken at the start of the trial.

The EPG, OPG, PCV, and FAMACHA© data were analyzed by repeated measures analysis in a completely randomized design using the mixed model procedure of SAS (SAS, 2008). Treatment, period (sampling date), and the treatment x period interaction were included in the model. The EPG and OPG data were log-transformed [ln(EPG + 1), ln(OPG + 1)] prior to statistical analysis. These data were reported as least squared means, with statistical inferences based upon log-transformed data analysis.

Results and Discussion: Both SL hay and SL silage diets reduced GIN and coccidial egg excretion (EPG and OPG, figures 1 and 2) in young kids compared with goats on the control (BG hay) diet. Treatment ($P < 0.02$), period ($P < 0.001$), and treatment x period interaction ($P < 0.03$) effects were all significant for EPG data, while treatment and period ($P < 0.001$) effects were significant for OPG data. For the goats on the SL hay treatment diet, there were significant reductions in EPG counts on Day 14 ($P < 0.001$) and Day 21 ($P < 0.01$), with a trend for lower EPG on Day 28 ($P = 0.082$). The SL hay-fed kids had lower OPG than control animals (BG hay diet) on Day 7 ($P < 0.02$) and Days 14, 21, and 28 ($P < 0.001$). These results with SL hay confirm previous reports of the anti-parasitic properties of dried SL for both sheep (Lange et al., 2006) and goats (Shaik et al., 2006). In a study with unground (long) SL and BG hays fed to young goats at 70% of the diet, Shaik et al. (2006) reported an 80% reduction in FEC in SL-fed animals after 7 days, with these differences maintained to the end of the experiment. Other studies have shown similar results for SL hay fed to goats in ground form (Terrill et al., 2009),

SL leaf meal (Terrill et al., 2008), and SL whole plant and leaf only pellets (Terrill et al., 2007; Kommuru et al., 2014), with initial significant differences occurring at between 7 to 28 days post feeding. In the current study, reductions in FEC relative to control in goats fed SL hay ranged from 75% to 85% (Figure 1). The reductions in coccidial OPG in SL hay-fed goats (86% to 91%) confirm results from feeding trials with SL pellets for both sheep (Burke et al., 2013) and goats (Kommuru et al., 2014).

There was a differential effect of the SL silage diet on goat parasites, with a greater effect on coccidia (OPG; 79% to 84% reduction relative to control) than GIN (EPG; 57% to 72% recuction relative to control). However, there was a significant reduction in EPG in kids on the SL silage diet compared with the control diet on Day 14 ($P < 0.02$) and a trend for reduced EPG on Day 21 ($P = 0.058$). Feeding an ensiled SL diet to the kids significantly reduced their OPG values relative to control on Day 14 ($P < 0.02$), Day 21 ($P < 0.02$), and Day 28 ($P < 0.03$). These results with ensiled SL compare favorably with other reports on the anti-parasitic potential of ensiled CT-containing legumes. Heckendorn et al. (2006) reported 58% and 48% reductions in *H. contortus* EPG in young lambs fed sainfoin (*Onobrychis viciifolia*) hay and silage, respectively, compared with a non-tannin control forage.

There were no treatment effects on blood PCV or FAMACHA© scores in the kids, with treatment group averages ranging from 19.5 to 20.7 for PCV and 2.1 to 2.4 for FAMACHA© scores. This is likely due to the short duration of this study (28 days), as a number of authors have reported a lag in changes in anemia indicators (PCV and FAMACHA©) compared with egg excretion reduction due to the effect of SL in the diet of sheep and goats (Terrill et al., 2012).

Based upon these results, ensiling of SL did not appear to affect its anti-parasitic effectiveness against either GIN or coccidia. The small differences in efficacy between SL hay and SL silage may have been related to intake of each diet by the goats. The kids immediately consumed their ration of SL hay, but took a few days to get used to the SL silage. After the first week of the trial, both SL treatment groups consumed all of their rations. In this trial, all of the leaves were collected after cutting of fresh SL for both the hay and silage treatments. Normally, leaf loss during baling can be a problem when making SL hay due to the very rapid drying of SL leaves compared to the stems. Cutting and ensiling SL at a higher moisture level may be an effective way to capture more leaves while maintaining the anti-parasitic efficacy of this crop.

Literature Cited:

Burke, J.M., J.E. Miller, J.A. Mosjidis, and T.H. Terrill. 2012a. Grazing sericea lespedeza for control of gastrointestinal nematodes in lambs. Vet. Parasitol. 186:507-512.

Burke, J.M., J.E. Miller, J.A. Mosjidis, and T.H. Terrill. 2012b. Use of a mixed sericea lespedeza and grass pasture system for control of gastrointestinal nematodes in lambs and kids. Vet. Parasitol. 186:328-336.

Burke, J.M., J.E. Miller, T.H. Terrill, S.T. Orlik, M. Acharya, J.J. Garza, and J.A. Mosjidis. 2013. Sericea lespedeza as an aid in the control of *Eimeria* spp. in lambs. Vet. Parasitol. 193:39-46.

Heckendorn, F., D.A. Haring, V. Maurer, J. Zinsstag, W. Langhans, and H. Hertzberg. 2006. Effect of sainfoin (*Onobrychis viciifolia*) silage and hay on established populations of *Haemonchus contortus* and *Cooperia curticei* in lambs. Vet. Parasitol. 142:293-300.

Kommuru, D.S., T. Barker, S. Desai, J.M. Burke, A. Ramsay, I. Mueller-Harvey, N. Kamisetti, and T.H. Terrill. 2014. Use of pelleted sericea lespedeza (*Lespedeza cuneata*) for natural control of coccidia and gastrointestinal nematodes in weaned goats. Vet. Parasitol. 204:191-198.

Lange, K.C., D.D. Olcott, J.E. Miller, J.A. Mosjidis, T.H. Terrill, J.M. Burke, and M.T. Kearney. 2006. Effect of sericea lespedeza (*Lespedeza cuneata*) fed as hay, on natural and experimental *Haemonchus contortus* infections in lambs. Vet. Parasitol. 141:273-278.

Min, B.R., W.E. Pomroy, S.P. Hart, and T. Sahlu. 2004. The effect of short-term consumption of a forage containing condensed tannins on gastro-intestinal nematode parasite infections in grazing wether goats. Small Rumin. Res. 51:279-283.

SAS Institute, 2008. SAS/STAT Software: changes and enhancements.Release 9.2, SAS Technical Report. SAS Institute, Cary, NC.

Shaik, S.A., T.H. Terrill, J.E. Miller, B. Kouakou, G. Kannan, R.M. Kaplan, J.M. Burke, and J.A. Mosjidis. 2006. Sericea lespedeza hay as a natural deworming agent against gastrointestinal nematode infection in goats. Vet. Parasitol. 139:150-157.

Terrill, T.H., J.A. Mosjidis, D.A. Moore, S.A. Shaik, J.E. Miller, J.M. Burke, J.P. Muir, and R.Wolfe. 2007. Effect of pelleting on efficacy of sericea lespedeza hay as a natural dewormer in goats. Vet. Parasitol. 146:117-122.

Terrill, T.H., F.A. Soli, M. Vanguru, A. Vuggam, S.A. Shaik, J.A. Mosjidis, J.E. Miller, B. Kouakou, and J.M. Burke. 2008. Effectiveness of sericea lespedeza leaf meal to reduce worm burden in goats. 9th Int. Conf. on Goats, August 31st to September 5th, 2008, Queretaro, Mexico, p. 258.

Terrill, T.H., G.S. Dykes, S.A. Shaik, J.E. Miller, B. Kouakou, G. Kannan, J.M. Burke, and J.A. Mosjidis. 2009. Efficacy of sericea lespedeza hay as a natural dewormer in goats: Dose titration study. Vet. Parasitol. 163:52-56.

Terrill, T.H., J.E. Miller, J.M. Burke, J.A. Mosjidis, and R.A. Kaplan. Experiences with integrated concepts for the control of *Haemonchus contortus* in sheep and goats in the United States. Vet. Parasitol. 186:28-37.

Van Wyk, J.A., and G.F. Bath. 2002. The FAMACHA© system for managing haemonchosis in sheep and goats by clinically identifying individual animals for treatment. Vet. Res. 33:509–529.

Whitlock, H.V. 1948. Some modifications of the McMaster Helminth egg counting technique and apparatus. J. Coun. Sci. Ind. Res. 21:177-180.

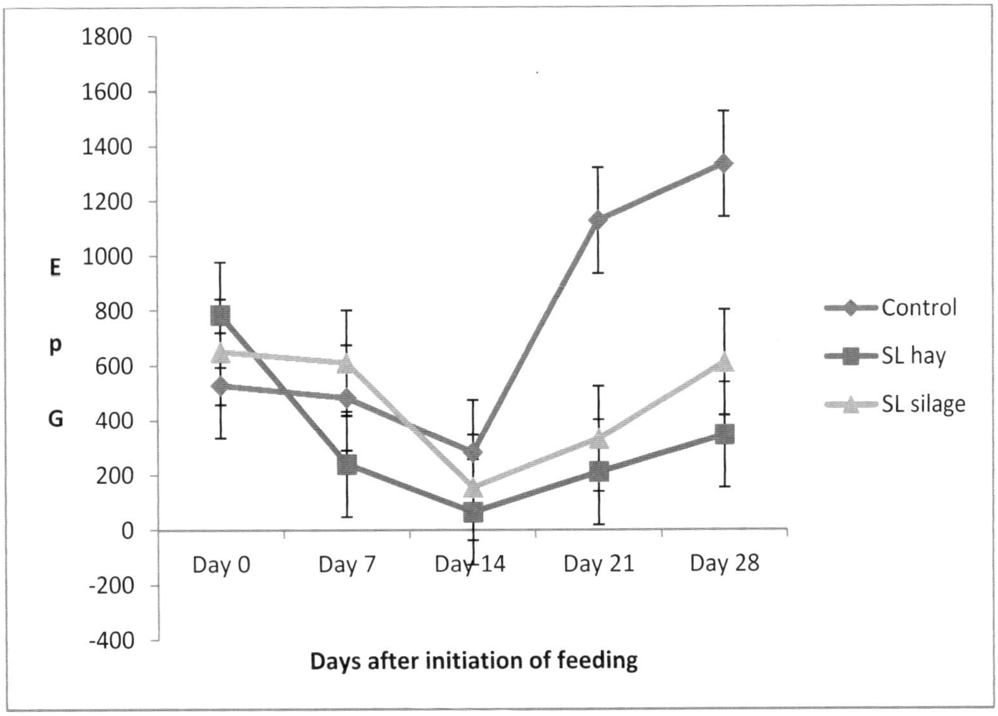

Figure 1. Least squares means of gastrointestinal nematode egg counts (eggs/g, EPG) in feces of Spanish kids fed sericea lespedeza (SL) hay, SL silage, or bermudagrass hay (control) diets.

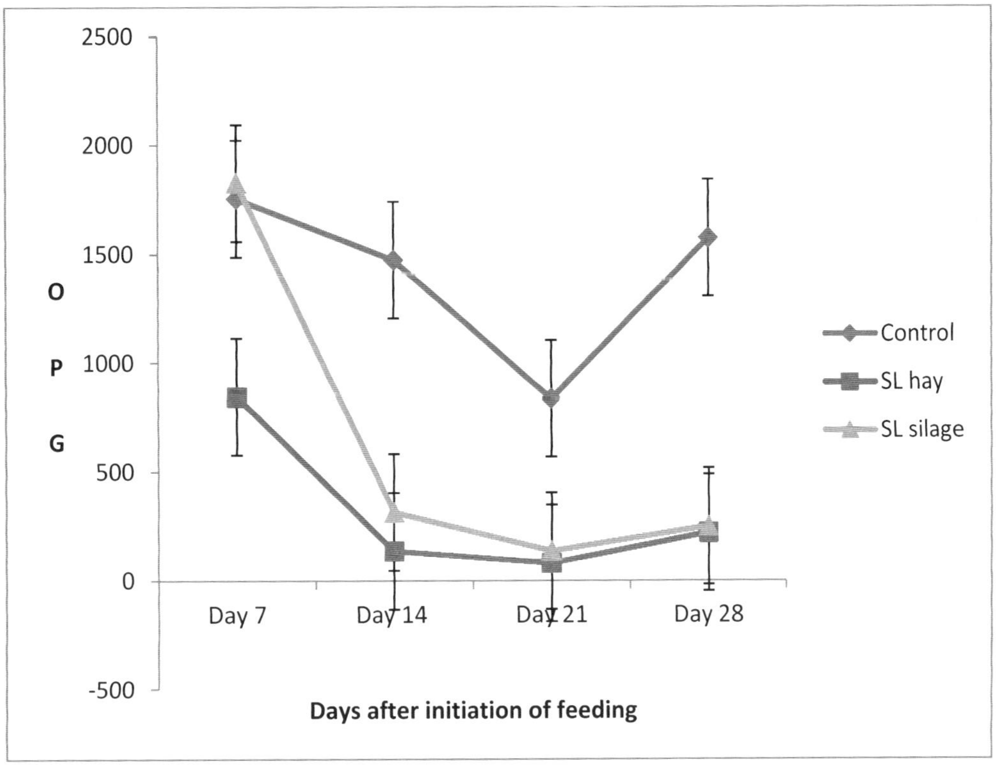

Figure 2. Least squares means of coccidial oocyst counts (oocysts/g, OPG) in feces of Spanish kids fed sericea lespedeza (SL) hay, SL silage, or bermudagrass hay (control) diets.

PERFORMANCE OF BERMUDAGRASS CULTIVARS AND ECOTYPES IN THE NORTHERN TRANSITION ZONE OF THE UNITED STATES

C.D. Teutsch, K.M. Mercier, and K. Quick[1]

Bermudagrass (Cynodon dactylon L.) could provide summer grazing for ruminant livestock in Virginia and other transition zone states. Common bermudagrass ecotypes exist on many farms in the Piedmont and Coastal Plains regions of Virginia. Little is known about the performance of these ecotypes under improved management. The objective of this study was to determine the performance of hybrid, seeded, and common ecotypes of bermudagrass under improved management. In June 2012, 5 common ecotypes, 24 seeded, and 1 hybrid bermudagrass were established at Virginia Tech's Southern Piedmont AREC, Blackstone, VA. The experimental design was a randomized complete block with four replications. Yield was determined by clipping a 5 ft strip through the center of each plot using sickle bar-type, self propelled forage harvester. Due to heavy weed pressure during establishment, data for 2012 and 2013 are not presented. In 2014, ecotypes and the hybrid control yielded more than the seeded types (11,708 and 11,435 lb/A versus 10,181 lb/A). In 2015, there were no differences between the three types of bermudagrass ($P = 0.24$) with yields ranging from 15342 to 16682. These data indicate that well-adapted local ecotypes of bermudagrass could be managed to meet the summer grazing needs of livestock producers in the northern transition zone. More work is needed to determine the yield stability of these ecotypes in other geographic areas within the transition zone.

[1]Associate Professor and Graduate Research Assistants, Virginia Tech's Southern Piedmont Agriculture Research and Extension Center, Blackstone, VA 23824.

SELECTING SUMMER ANNUAL CULTIVARS: IT IS ABOUT MORE THAN JUST YIELD!

C.D. Teutsch, K.M. Mercier, and K. Quick[1]

Summer annual grasses can provide high quality summer grazing for ruminant livestock in Virginia and other transition zone states. In the past, recommendations for cultivar selection have been to "find a reasonably priced, locally available variety and focus on management". While good agronomic and grazing management are crucial when using summer annual grasses, recent advances in breeding may change how varieties are selected. The impact of the brown midrib (BMR) trait on digestibility and the relationship between yield and digestibility was examined for five years of data from summer variety annual trails conducted at Virginia Tech's Southern Piedmont AREC, Blackstone, VA. Cultivars possessing the BMR trait were approximately 5% more digestible on average. In 2009 and 2010, no relationship between yield and digestibility was found. In 2011, 2012, and 2014, a negative relationship between yield and digestibility occurred. When yield and digestibility were indexed against the average yield and digestibility for a given trial and graphed, varieties with above average yield and digestibility could be visualized. These are cultivars that should be considered for forage programs. Cultivars that had above average yield and digestibility for three out of five years included AS6501, AS9301, and Xtragraze BMR. This research indicates that both yield and digestibility should be considered when selecting summer annual grasses for forage programs.

[1]Associate Professor and Graduate Research Assistants, Virginia Tech's Southern Piedmont Agriculture Research and Extension Center, Blackstone, VA 23824.

STOCKPILING NOVEL ENDOPHYTE TALL FECUE FOR SUMMER GRAZING

C.D. Teutsch, B.T. Campbell, N.T. Shireman, K.M. Mercier, and K.K. Quick[1]

In the transition zone of the U.S., high temperatures and drought often limit forage availability during the summer months. This area of the U.S. is also dominated by tall fescue, most of which is infected with the toxic endophyte. The use of tall fescue pastures during the summer months has been discouraged due to tall fescue toxicois. The incorporation of novel endophyte tall fescue into grazing systems has the potential to alter usage patterns. Past research has focused on stockpiling tall fescue in late summer for winter grazing. However, non-toxic tall fescue could potentially be stockpiled during spring to provide grazing during the summer. The objective of this study was to evaluate animal performance on novel endophyte tall fescue stockpiled for summer grazing. Treatments were 1) tall fescue growth allowed to accumulate from spring green up and 2) tall fescue clipped in mid-May and growth allowed to accumulate. All pastures were fertilized with 60 lbs of N/acre in mid-May. Heifers in 2014 (551 lb avg.) and steers in 2015 (491 lb avg.) were randomly assigned to the treatments. Pasture biomass and nutritive value were monitored biweekly. Cattle were weighed every two weeks and average daily gain (ADG) was calculated. Heifers grazing pastures that were clipped had higher ADG's than calves grazing unclipped pastures in 2014 (1.38 versus 1.20 lb/day; $P = 0.05$) and in 2015 (1.40 versus 1.09 lb/day; $P = 0.09$). Novel endophyte tall fescue stockpiled for summer grazing can support reasonable levels of ADG, especially if pastures are clipped or grazed to remove reproductive tissue prior to stockpiling.

[1]Associate Professor, Virginia Tech, Southern Piedmont AREC, Blackstone, VA (cteutsch@vt.edu); Technical Support Manager, DSM Nutritional Products, (Brian-T.Campbell@dsm.com); Graduate Students, Virginia Tech, Southern Piedmont AREC, Blackstone, VA (nathan8@vt.edu, kmerc094@vt.edu, kkquick@vt.edu).

NITROGEN AND LEGUME MANAGEMENT IN SOUTHEASTERN WILDRYE

[1]Matthew Thornton and [2]J. Brett Rushing

Abstract

Southeastern wildrye (*Elymus glabriflorus*) is a native, cool-season, perennial grass that, until recently, has been relatively un-researched. In order to validate and quantify this species' potential as a forage crop, applied agronomic studies are required. Field trials were established in the fall of 2014 to compare annual legumes and synthetic nitrogen applications in southeastern wildrye stands. Locations for this study include Starkville and Newton, Mississippi. The study was designed in a randomized complete block with four replications. Five annual clover treatments (arrowleaf, balansa, ball, berseem, and crimson clover) and five nitrogen treatments (0, 50, 100, 150, and 200 lbs total N/acre) were evaluated. Nitrogen was applied in split applications following each harvest. Data collection included plant heights, botanical composition, botanical ratings, normalized differential vegetation index (NDVI; nitrogen treatments only), biomass yields, and forage quality analysis. Forage quality analysis included acid detergent fiber (ADF), neutral detergent fiber (NDF), crude protein (CP), and total digestible nutrients (TDN). Also, nitrogen use efficiency was calculated for nitrogen treatments. Following one year of data collection (three harvests), southeastern wildrye was found responsive to the higher rates of applied nitrogen, with dry matter (DM) yields ranging from 1210 to 5173 lbs/acre DM across treatments. However, annual clover treatments produced similar DM yields with superior nutritional characteristics, including CP (15-21%), lower ADF (27-32), and lower NDF (38-53), making them a viable alternative to nitrogen fertilizer.

Introduction: In Mississippi, tall fescue (*Schedonorus arundinaceus*) is a wide spread hearty, perennial, grass that provides the base of many forage systems. Due to it widespread adaptations and excellent forage characteristics it has been established on nearly 175,000 acres of pasture land in the state (Lemus 2015). 'Kentucky-31', a variety containing the toxic endophyte (*Neotyphodium coenophialum*), is the most commonly established variety and is found on the majority of those acres. Tall fescue plants infected with this endophyte benefit from greater persistence, greater yield during droughty or stressful periods (Bouton et al. 1993), increased seed germination, greater seedling vigor, greater tillering, tolerance of close grazing, and resistance to diseases and pests. The presence of *Neotyphodium coenophialum*, which produces an ergot alkaloid (Ergovaline) toxic to mammals, has negative effects on cattle resulting in lower performance than chemical forage analysis would suggest (Ball et al. 2007).

Southeastern wildrye (*Elymus glabriflorus*) is a native, perennial, cool-season grass species found throughout the eastern United States sharing portions of its range with the closely related Canada and Virginia wildrye (*E. canadensis* and *E. virginicus*), both of which it is planted in mixes with for soil stabilization. Southeastern wildrye is relatively tall and robust grass reaching 4ft at maturity. Tolerant of a wide variety of environmental conditions, it can be found growing on dry to moist sites, in full sun to partial shade, and neutral to slightly acidic soils that may be fine to coarse in texture.

[1]Unlike the more commonly planted native warm-season grass species, southeastern wildrye has a relatively high innate germination rate (70%) without stratification (Belt et al. 2013). In addition to a wide range of climatic adaptations and tolerance to full sun, Rushing and Baldwin (2013) determined it to be the most likely candidate for a native cool-season forage species because of its innately decent forage quality characteristics: 11.1 – 18% crude protein (CP), 24.7 – 35.7% acid detergent fiber (ADF), and 47.3 – 59.1% neutral detergent fiber (NDF).

Nitrogen (N) use in any agronomic system is crucial to the productivity of that crop. In forage systems the addition of mineral N fertilizers increases the vegetative growth and overall DM yield (Chalmers et al. 1990), resulting in greater ADG (Whitehead 1995). However, these N based fertilizers are often expensive to apply and can contribute to environmental pollution through runoff. This need for the reduction of input cost, as well as reducing excessive runoff resulting in air, drinking water, and ocean water pollution (Sherman and Gurwick 2009), has led to greater interest in utilizing legumes. Leguminous species form symbiotic relationships with N fixing *Rhizobium spp.* bacteria allowing them to take up atmospheric N (Oberson et al., 2013). While this N is relatively slowly released to the grasses, a perennial forage system would benefit most over multiple years of legume incorporation. Additionally, incorporation of legumes in a forage stand has been shown to increase cumulative DM yields and the overall forage quality of the stand (Sleugh et al. 2000) resulting in a potentially more profitable system.

The primary goal of this study is to assess the overall use of southeastern wildrye in a forage system, and to determine if its productivity in a low input system (addition of legumes to a monotypic grass stand) is similar to a high input system receiving nitrogen fertilizers. Understanding these characteristics will aid in directing research efforts and management strategies focusing on southeastern wildrye as a forage crop.

Materials and Methods: Field trials were established in Starkville and Newton, MS on 7 October, 2014 at the H.H. Leveck Animal Research Center on the campus of Mississippi State University and the Coastal Plain Branch Experiment Station, respectively. The soils at the Starkville site are fine, smectitic, thermic Fluvaquentic Hapludolls that belong to the specific soil pedon Catapla silty clay loam. These soils are largely cultivated and occasionally flooded. The soils at the Newton site are coarse-loamy, siliceous, semiactive, thermic Glossic Fragiudults that are primarily used in pasture. The Newton soils belong to the specific soil pedon Prentiss loam. This field trial is a randomized complete block design (RCB) with four replications. Treatments were applied to 6 ft x 10 ft plots. This trial is a five x five factorial arrangement of treatments consisting of annual clovers and nitrogen applications.

---

[1] [1]Graduant Assistant; Department of Plant and Soil Sciences; Mississippi State University

[2]Assistant Research/Extension Professor; Coastal Plain Branch Experiment Station; Mississippi State University

All plots were planted using an Almaco© (Nevada, IA) 8-row light duty grass seed drill with 8 inch row spacing and a rate of 20 pounds per acre pure live seed (PLS) of southeastern wildrye. Clover species planted include: arrowleaf clover (*Trifolium vesiculosum*), balansa clover (*T. michelianum*), ball clover (*T. nigrescens*), berseem clover (*T. alexandrinum*) and crimson clover (*T. incarnatum*). All clovers were planted at PLS simultaneously with southeastern wildrye at a rate of arrowleaf 5 lbs/acre, balansa 3 lbs/acre, ball 2 lbs/acre, berseem 10 lbs/acre, and crimson 10 lbs/acre. Fertilizer was applied as urea/ammonium sulfate (33-0-0-S) in one-third split applications on 10 Feb (Starkville) and 11 Feb (Newton) then following harvest on Apr 8 and May 3 2015 in Starkville and on Apr 2 and May 1 2015 in Newton. Yearly total N rates include: 0, 50, 100, 150, and 200 pounds per acre. Prior to each harvest, sward heights were measured using a rising plate meter (Jenquip, New Zealand). Also, normalized differential vegetation index (NDVI) readings were taken in all nitrogen treatment plots using a handheld Greenseeker (Timble, Westminster, CO). To determine species contribution within each plot, southeastern wildrye, weeds, and clovers were hand separated from a random 14" x 14" in subplot. Biomass harvests were conducted using a Wintersteiger Cibus F plot harvester (Wintersteiger AG, Austria) approximately every 30 d, or when plot heights reached 12 inches in height. Harvest dates for Starkville include Apr 8, May 3, and Jun 4, whereas harvest dates for Newton were Apr 2, May 1, and Jun 5. All plots were clipped to a height of 4.5 inches. Subsamples were taken from each plot to determine percent DM. Each sample was dried in a forced air over at 122°F until no other change in moisture was observed. Dried samples were then ground through a 0.04 inch mesh screen using a Wiley-type mill (Thomas Scientific, Swedesboro, NJ). Ground samples were then analyzed for acid detergent fiber (ADF), neutral detergent fiber (NDF), and crude protein (CP) using near infrared reflectance spectroscopy (NIRS) analysis based off the mixed grass hay equation (NIRS Consortium, Hillsboro, WI). Total digestible nutrients (TDN) was calculated for all treatments and harvests. Nitrogen use efficiency was also calculated for N treated plots. Statistical analysis was conducted using ANOVA in SAS (SAS Institute, Cary, North Carolina). PROC GLM was used to determine least significant differences (LSD). Mean separations were based on Tukey's protected LSD and differences were considered significant at 0.05 probability level.

Results and Discussion: For the sake of brevity, only forage quality (ADF, NDF, CP, and TDN) and quantity (DM yields) results will be discussed. Differences were present between locations and between treatments at each location for both DM yields and forage quality analysis (Table 1 and 2; $P<0.05$). In Starkville, the berseem treatment (5364.1 lb/a) generated the greatest cumulative DM yield, while the 150 lb N treatment (5172.7 lb/a) had the greatest cumulative DM yield in Newton. For both locations, DM yields were similar between clover and N treatments. As the season progressed, frequent harvesting combined with increasing temperatures weakened clover stands, resulting in greater percentage of grass within each plot (data not shown).

Significant interactions were observed for ADF, NDF, CP, and TDN at each location within each harvest (Table 1 and 2). In general, the higher N (150 and 200 lbs N/acre) treatments performed similarly to the clover treatments. However, on average across all harvests, collective differences between clover and N treatments were minimal.

For example, mean TDN across all harvests for N treatments was 58.3%, versus 58.1% for all clover treatments at the Starkville location. Similarly, at the Newton location, mean TDN across all harvests for N treatments was 59.9%, versus 59.6% for all clover treatments. One point to note is the relatively high innate quality characteristics of southeastern wildrye. For the 0 lbs/acre N treatment, CP values remained above 14%, and > 57% TDN throughout the duration of the trial.

Analysis of one year's data indicates southeastern wildrye produces moderate amounts of nutritious DM, whether supplemented with chemical nitrogen or annual clovers. From an economic standpoint, however, sowing annual clovers into a stand of southeastern wildrye will enhance forage quality as a less expensive practice compared to applying synthetic N, especially during the earlier part of the growing season (April and May). As the season progresses, it may prove beneficial to supplement forage stands with N to generate more DM prior to seed development and maturation. In terms of quality, however, southeastern wildrye produces excellent values with or without added N or clover inclusion. This is an excellent attribute for low input grazing systems. A second growing season is required to confirm the results gained from the first year of data collection. Information gathered from this project will help validate the need for future development and commercialization of southeastern wildrye as a potential forage crop for Mississippi.

Literature Cited:

Ball, D.M., Hoveland, C.S., Lacefield, G.D., 2007. Southern Forages. p. 199-205. 4th ed. International Plant Nutrition Institute. Norcross, GA.

Belt, S., B. Rushing, S. Tangren. 2013. Plant Guide for Southeastern wildrye (*Elymus glabriflorus*). USDA-NRCS, Norman A. Berg National Plant Materials Center. Beltsville, MD 20705.

Bouton, J.H., R.N. Gates, D.P. Belesky, and M. Owsley. 1993. Yield and persistence of tall fescue in the southeastern coastal plain after removal of its endophyte. Agron. J. 85:52-55.

Chalmers, A.G., C.D. Kershaw, and P.K. Leech. 1990. Survey of Fertilizer Practice: Fertilizer Use on Farm Crops in England and Wales, 1986. MAFF, London.

Lemus, R. 2015. Tall fescue persistence in Mississippi. Mississippi State University Extension Service: Forage News Vol. 8.

Oberson, A., E. Frossard, C. Buhlmann, J. Mayer, P. Mader, and A. Luscher. Nitrogen fixation and transfer in grass-clover leys under organic and conventional cropping systems. 2013. Plant Soil 371: 237 – 255.

Rushing, J.B. and B.S. Baldwin. 2013. Evaluation of wildrye (Elymus spp.) as a potential forage and conservation planting for northeast Mississippi. Native Plants Journal. 14 (3):192-03.

Sherman, D.G. and N. Gurwick. 2009. No sure fix. Union of Concerned Scientist. UCS Publications. Cambridge, MA.

Sleugh, B., K.J. Moore, J.R. George, and E.C. Brummer. 2000. Binary legume-grass mixtures improve forage yield, quality, and season distribution. Agron. J. 92:24-29.

Whitehead, D.C. 1995. Grassland Nitrogen. CAB International. Wallingford, Oxon, UK.

Table 1. Forage quality analysis and dry matter (DM) yields for southeastern wildrye with nitrogen and clover treatments at the Starkville, MS location during spring of 2014.

Treatment	ADF (%)			NDF (%)			CP (%)			TDN (%)			DM Yield (lbs/acre)
	Apr 8	May 3	Jun 4	Apr 8	May 3	Jun 4	Apr 8	May 3	Jun 4	Apr 8	May 3	Jun 4	
Arrowleaf	33.8 ab[1]	31.7 b	31.9 bcd	50.1 ab	47.1 bcd	48.7 b	16.3 cd	16.6 cdef	18.3 cde	57.0 cd	58.6 b	58.4 bcd	1576.3 bc
Balansa	30.7 cd	31.1 b	30.7 cde	45.7 c	43.6 cde	50.7 ab	20.9 a	17.7 bc	18.6 cd	59.3 ab	59.0 b	59.3 abc	1600.7 bc
Ball	30.5 d	28.0 c	32.2 bc	48.4 bc	42.2 de	50.0 b	17.6 bc	15.7 ef	17.7 de	59.5 a	61.3 a	58.2 cd	1338.7 bc
Berseem	35.1 a	31.1 b	35.2 a	49.4 ab	40.4 e	44.3 c	21.1 a	21.1 a	19.5 bc	56.0 d	59.0 b	56.0 e	5364.1 a
Crimson	35.3 a	34.7 a	32.4 bc	50.1 ab	48.0 abc	53.1 a	14.5 d	16.7 cde	17.5 de	55.8 d	56.3 c	58.0 cd	1245.3 c
0 lbs N/a	35.3 a	33.3 ab	33.5 ab	52.3 a	48.7 abc	53.0 a	16.4 cd	16.0 def	16.8 e	55.9 d	57.3 cb	57.2 de	1930.4 bc
50 lbs N/a	30.7 cd	32.4 ab	32.2 bc	49.3 ab	52.3 a	53.3 a	15.9 cd	15.4 f	18.1 cde	59.3 ab	58.0 cb	58.2 cd	1210.6 c
100 lbs N/a	32.4 bcd	32.2 b	31.4 cd	50.5 ab	51.6 ab	51.6 ab	17.7 bc	15.9 ef	19.7 bc	58.1 abc	58.2 b	58.7 bc	2523.1 bc
150 lbs N/a	31.9 bcd	31.2 b	29.6 e	50.6 ab	51.0 ab	49.8 b	18.0 bc	17.2 cd	21.0 ab	58.4 abc	58.9 b	60.1 a	1751.7 bc
200 lbs N/a	33.0 abc	31.2 b	30.3 de	51.0 ab	48.7 abc	48.7 b	19.5 ab	18.7 b	21.7 a	57.6 bcd	58.9 b	59.6 ab	2733.5 b

[1]Lower cased letter denotes significant differences using Tukey's protected LSD with all differences considered significant at $\alpha = 0.05$.

Table 2. Forage quality analysis and dry matter (DM) yields for southeastern wildrye with nitrogen and clover treatments at the Newton, MS location during spring of 2014.

Treatment	ADF (%)			NDF (%)			CP (%)			TDN (%)			DM Yield (lbs/acre)
	Apr 2	May 1	Jun 5	Apr 2	May 1	Jun 5	Apr 2	May 1	Jun 5	Apr 2	May 1	Jun 5	
Arrowleaf	29.2 a[1]	29.2 c	33.9 ab	45.1 ab	46.4 b	52.9 bcd	14.8 e	15.4 def	16.8 bc	60.5 c	60.5 b	56.9 bc	2533.1 bcd
Balansa	26.2 c	27.7 d	33.6 ab	38.3 c	39.1 d	53.8 abc	20.6 a	18.9 ab	16.9 bc	62.7 a	61.5 a	57.1 bc	3018.4 bcd
Ball	28.0 ab	27.7 d	32.5 abc	42.6 b	42.0 c	51.6 cd	16.7 cde	15.4 def	17.7 bc	61.4 bc	61.6 a	58.0 abc	2929.3 bcd
Berseem	29.2 a	29.4 bc	34.8 a	43.1 b	41.8 c	49.5 d	19.6 ab	19.2 a	18.4 ab	60.5 c	60.3 bc	56.2 c	3660.4 b
Crimson	28.6 a	32.0 a	34.3 ab	44.2 ab	49.7 a	55.8 ab	15.6 de	14.4 ef	16.5 c	60.9 c	58.3 d	56.6 bc	2221.4 cd
0 lbs N/a	28.9 a	30.6 b	32.6 abc	45.4 ab	49.7 a	51.6 cd	14.4 e	14.2 f	16.8 bc	60.7 c	59.4 c	57.9 abc	1945.9 d
50 lbs N/a	28.8 a	29.6 bc	35.1 a	46.3 a	48.2 ab	57.3 a	15.6 de	15.8 de	16.6 c	60.7 c	60.1 bc	56.0 c	3359.2 bc
100 lbs N/a	26.8 bc	29.4 bc	32.0 bc	44.1 ab	48.8 ab	52.7 bcd	17.8 bcd	15.9 de	17.9 abc	62.3 ab	60.3 bc	58.3 ab	3477.6 b
150 lbs N/a	26.8 bc	29.3 bc	33.9 ab	44.3 ab	49.3 a	56.0 ab	19.5 ab	16.4 cd	17.9 abc	62.2 ab	60.4 b	56.9 bc	5172.7 a
200 lbs N/a	26.3 c	28.9 cd	30.4 c	44.4 ab	50.0 a	51.8 cd	18.7 abc	17.5 bc	19.4 a	62.6 a	60.7 ab	59.6 a	3299.3 bc

[1]Lower cased letter denotes significant differences using Tukey's protected LSD with all differences considered significant at $\alpha = 0.05$.

EVALUATION OF PRE- AND EARLY POST-EMERGENT APPLICATIONS OF IMAZAPIC TO SOUTHEASTERN WILDRYE

M. Thornton[1] and J. B. Rushing[2]

Southeastern wildrye (*Elymus glabriflorus*) is a native, perennial, bunch-type grass that grows in a wide variety of soils and climates throughout the eastern United States. In order for this species to be considered a viable forage crop, basic agronomic research must be conducted to define its capabilities. One crucial step in this process is determining establishment protocols, specifically herbicide tolerance. Imazapic {(±)-2-[4,5-dihydro-4-methyl-4-(1-methylethyl)-5-oxo-1*H*-imidazol-2-yl]-5-methyl-3-pyridinecarboxylic acid}, a commonly used herbicide in perennial native species, is a likely candidate for weed control in southeastern wildrye. Field trials were established in Starkville and Newton, Mississippi in fall of 2014 to evaluate plant response and weed control efficacy. Treatments were arranged in a randomized complete block design containing 0, 2, 4, 6, and 8 oz/acre rates of imazapic in both pre-emergent (at planting) and early post-emergent (late winter) application timings. Data collection included plant heights and visual ratings at 2, 4, 6, and 8 weeks following application of herbicide. At full seed maturity, a final harvest was conducted to determine dry matter (DM) and seed yield. Preliminary data suggest that lower rates (2 and 4 oz/acre) of imazapic applied at pre-emergence is beneficial in controlling winter and early spring annual weeds. Southeastern wildrye also displays tolerance to imazapic at these rates. Higher rates of imazapic, particularly those applied early post-emergent, resulted in stunting which can be particularly damaging to a stand as weed competition in early spring can be strong. A second year of testing is underway to further validate these findings.

[1]Graduate Student; Department of Plant and Soil Sciences; Mississippi State University
[2]Assistant Research/Extension Professor; Coastal Plain Branch Experiment Station; Mississippi State University

ARROWLEAF SIDA CONTROL IN PERENNIAL WARM-SEASON GRASS PASTURES

E.K. Twidwell, R.E. Strahan, S.J. Dutile and M.J. Simon[1]

Arrowleaf sida (*Sida rhombifolia* L.) is a summer annual broadleaf weed species that infests many perennial warm-season grass pastures throughout the southeast region of the USA. It is a weed species that is not palatable for livestock and its aggressive nature competes with forage grasses for light, water and nutrients. This study was conducted to compare six broadleaf pasture herbicides for control of arrowleaf sida in a bahiagrass pasture. The herbicides were applied on June 30, 2014. Arrowleaf sida control ranged from 76 to 95% at 30 days after treatment, with four of the herbicides providing over 90% control. By October 30, the herbicides Weedmaster, Grazon P+D and GrazonNext HL still provided over 80% control. The same herbicides were re-applied to the same plot area on July 2, 2015. Results obtained were very similar to those found in 2014. On October 1, arrowleaf sida control ranged from 30 to 86%. The herbicides Weedmaster, Grazon P+D and GrazonNext HL provided over 80% control while the herbicides 2,4-D, PastureGard HL and Crossbow provided relatively poor control. Results suggest that arrowleaf sida is a relatively difficult broadleaf weed species to control, and it is important for producers to select the appropriate herbicide that will provide acceptable control.

[1]Proessor, LSU AgCenter, Baton Rouge, LA (etwidwell@agcenter.lsu.edu); Associate Professor, LSU AgCenter, Baton Rouge, LA (rstrahan@agcenter.lsu.edu); Extension Agent, LSU AgCenter, Lafayette, LA (sdutile@agcenter.lsu.edu); Farm Operations Manager, UL Cade Farm, St. Martinville, LA (Mjs0563@louisiana.edu).

EVALUATION OF ALFALFA HAY WHEN USING HAY PERSERVATIVES IN MISSISSIPPI

J. White and R. Lemus[1]

Alfalfa (*Medicago sativa* L.) production in Mississippi has increased in recent years due to the increase in research and demonstrations outlining proper management strategies. However, the process of curing and baling quality hay can prove to be difficult in a humid climate. The use of hay preservatives such as Hay-Guard could help mitigate "bale heating" observed while baling at higher moisture. A three-acre field of first year alfalfa in Starkville, Mississippi was utilized for the study taking three separate harvest on April 29, June 16, and July 27. Hay Guard was applied at a rate of 0, 2 and 4 lb/ton at baling. The hay was squared-baled when windrows tested between 25 and 21% moisture and again when tested at 20% moisture or below. A total of 48 bales were taken for each harvest and arranged in a randomized compete block design with 3 replications. From these bales, the following data was recorded: (1) weights were taken at 0, 5, 15, and 20 days; (2) core temperature was taken at 0, 4, 8, 12, 16, 20 days, and (3) hay cores were taken at 0, 10, and 20 days after bailing. Bale temperatures and forage quality (lignin and crude protein) were affected by harvest date and timing of sampling. The use of the preservative and the moisture at which the hay was baled had no effect on forage quality and bale temperature. This was most likely due to abnormally dry summer weather that made it problematic in baling higher moisture hay. Crude protein (21%) was greatest, and lignin (8%) and bale temperatures were lowest (78° F) in the first harvest.

[1]Variety Testing Manager, Plant and Soil Sciences, Mississippi State University, Starkville MS (jwhite@pss.msstate.edu); Forage Extension Specialist, Plant and Soil Sciences, Mississippi State University, Starkville MS. (RLemus@pss.msstate.edu).

AUTHOR INDEX

AUTHOR INDEX